Get It Right the First Time!

Get It Right the First Time!

PHARMACEUTICAL PRODUCT DEVELOPMENT

Intelligent Solutions for Challenging Issues

BERNARD D. KING, MD, MBA

Macnas Press
San Diego, CA

Macnas Press
4980 Beauchamp Ct.
San Diego, CA 92130 USA

LCCN: 2010903532
ISBN: 978-0-9845019-0-8

Publishers Cataloguing-in-Publication Data

King, Bernard D., 1949-

Get it right the first time! : pharmaceutical product development : intelligent solutions for challenging issues / Bernard D. King. --
San Diego, CA : Macnas Press, c2010.

p. ; cm.

ISBN: 978-0-9845019-0-8
Includes bibliographical references and index.

1. Drug development. 2. Medical instruments and apparatus--Development. 3. Pharmaceutical industry--Research. 4. Pharmaceutical technology. 5. Medical instruments and apparatus--Technological innovations. 6. Drug development-- Technological innovations. I. Title. II. Title: Pharmaceutical product development.

RM301.25 .K56 2010 2010903532
615/.19--dc22 1003

Printed in the United States of America

5 4 3 2 1

Table of Contents

Part III Regulating the Plan: Deliver the Best Possible Product to the Customer

Part IV Changing the Plan: Trust Me, The Future Will Be Different

Preface

The ultimate goal of all drug, biologic or device development is to deliver products to the marketplace that improve the health of people throughout the world. Achieving that goal requires vision, determination, persistence, execution and a bit of luck. To me, achieving that goal is the magic of this business and the reason to go to work each day.

Throughout this book a fundamental operating assumption is that, to achieve our goal, we are driving product development (of a drug, biologic, device, or combination product) to achieve a significant value event. The specific nature of that event will be determined by the particular situation of those responsible for development. The event might be a partnering event, a merger or acquisition, or delivering the drug to the marketplace after receiving marketing approval.

My background includes broad experience in the pharmaceutical, biotechnology and medical device industries. After a career as a practicing cardiologist, I have been in the industry for nearly 25 years. I started as a medical monitor and, over the years, have run research and development, have had responsibility for worldwide development, and have run companies. I have worked in very large multinational companies and small start-ups. For the last several years, my consulting company has worked with a broad range of clients around the world in all areas of development. All of this gives me a fairly unique perspective. That perspective will, I believe, enable us collectively to explore clinical development and place it squarely within sound operational and regulatory frameworks.

Product development in the health care industry is an intensely challenging, frustrating, exhilarating, complex, demanding process with odds of success that are miniscule at best and positively obscene at worst. Why, then, would anyone do this? For every product delivered to the market — even a small market, an orphan product, for example — the developer's dream is that a grateful patient will have access to the product and receive substantial benefit as a result of using it.

The development process for a product — whether the company involved is small or large, whether the product is destined to be a blockbuster or to only serve an orphan market, or whether activities are conducted in Europe, North America or anywhere else in the world — is fundamentally similar. However, within that "similarity" lay, on one hand, many potential opportunities for creative — yet ethical — development that can reward all involved handsomely while delivering the needed product to the market and, on the other hand, enormous pitfalls that can derail development, resulting in the loss of millions of dollars and NOT delivering that product to patients who need it.

Focus and Objectives

The focus of this book is to identify and explain the key components of the development process — from beginning to end — and identify the issues (challenges and opportunities) associated with each component that will impact the likelihood of achieving success.

Alice: *"Would you tell me, please, which way I ought to go from here?"*

Cheshire Cat: *"That depends a good deal on where you want to get to."*
—Lewis Carroll, "Alice in Wonderland"

If there is a guiding message throughout the book, it is — with full acknowledgement to Steven Covey — "Begin with the end in mind." (Covey 1990) I very strongly believe that if you identify where you are going, clarify your requirements for success, and then operate with those goals in mind, you ultimately have a much higher chance of success.

We will identify the key decision points and key issues that must be considered along the way. We will indicate points of relationship between regulatory and clinical disciplines. Given a constantly changing environment, we will offer some thoughts about maintaining a vibrant and robust development plan and process despite changes that may be entirely (or, at least, largely) outside your control.

Essentially all examples used in the book will apply particularly to drugs and biologics, but the principles and the approach will be absolutely applicable to devices.

We will be starting the book by framing development within the multiple contexts in which a company operates. All of these contexts influence, to one degree or another, the choices we make as we move forward in development. We will next move to a fairly detailed description of the product development process. Then, over subsequent sections, we will look at more operational issues.

A lot of material is covered in this book. I believe the volume and scope of material is necessary in order to identify and address the multitude of issues that impact clinical development. I make no assumptions about your background as you start this journey with me. In that sense, the book is self contained. However, even if you have substantial experience in clinical development, I believe there is sufficient material, and there are enough important questions raised, to further broaden and deepen your knowledge and insight.

We would love to have your feedback and thoughts about the topics discussed here and ways to improve the book. Please contact me at:

Barney King
Macnas Consulting International, Inc.
www.theclinicaltrialexpert.com

Covey, Stephen R. New York: Fireside, 1990.

Acknowledgements

The ideas expressed in this book have grown out of many years of working in the area of product development — pharmaceutical drugs, biotechnology products, and medical devices. During my nearly twenty-five years in the industry, first at SmithKline French (now GlaxoSmithKline), at Bristol-Myers Squibb (the ConvaTec division), several biotechnology companies (most notably Houghten Pharmaceuticals (Trega), Predict and Kemia), and at PAREXEL (a contract research organization), and then over the last decade consulting with many smaller and mid-sized companies from India to Israel, I have had the opportunity to work with some absolutely outstanding and dedicated individuals. While too many to mention here, those contributing the most to my thinking and development included Tom Beck, Mark Citron, Scott Gillis, Karen Church, Connie Crowley, Dan Kisner, and Mike Lack. What I have learned from this group of individuals regarding development, organization, planning, regulatory affairs, and business has been invaluable.

During the preparation of this book, my family has been most supportive of my time away and my time — although physically present — mentally, somewhere else. Kathleen, my wife, and my children, Matthew, Meaghan, Hillary, and Michael are sources of inspiration. Matthew also assisted with the indexing process and other textual suggestions.

I acknowledge that I unabashedly support the development of new medical products — drugs, biologics, devices, diagnostics, and combination products. These products make our health better, our daily activities more fulfilling, and our lives longer. While certainly not a perfect industry — none is — the ultimate aim of this industry is truly worthwhile. That moment a drug or device, to which your effort contributed, finally makes it to the marketplace and, in turn, betters the life of a patient, is indeed special.

Part I

Planning the Plan: Begin with The End in Mind

Chapter 1

Background

1.1 Historical Events in Drug Development

Although there were various efforts to regulate foods and drugs previously, what really got the whole process started in the United States was passage of the Food and Drug Act in 1906. (See Table 1-1.) This act created the Agency that eventually became today's Food and Drug Administration (FDA). This was followed, in 1938, by the Food, Drug and Cosmetic Act (FD&C Act), which is today really the guiding principle for all of us as we operate within the field. In particular, the FD&C Act introduced pre-market drug review in the form of a New Drug Application (NDA) that presented the sponsor's evidence that the new drug was safe.

Other particularly important developments have occurred since that time. In 1962, following the thalidomide tragedy, the Kefauver–Harris Amendments to the Act mandated an efficacy (in addition to safety) requirement for marketing approval. In 1985 and 1987, there were significant rewrites of the regulations for investigational new drug applications (INDs) and for NDAs; it was at that point that I came into the industry. That timing was fortuitous, since the guidances were freshly released at the time I was preparing and filing my first NDA.

The Orphan Drug Act of 1983 provided mechanism for expediting approval for drugs for rare diseases. In 1992, Accelerated Approval regulations were finally collected, codified and released. In the same year, the first PDUFA (the Prescription Drug User Fee Act) was also signed into law. This act authorized collection of fees for sponsors submitting applications for FDA review. Subsequently, PDUFA has been renewed every five years, but each renewal has been contentious.

Table 1.1. Historical Events in Drug Development

Event	Year
Pure Food and Drug Act	1906
Food, Drug and Cosmetic Act	1938
Durham-Humphrey Amendment	1951
Kefauver-Harris Amendments	1962
NDA Rewrite Regulations	1985
IND Rewrite Regulations	1987
Accelerated Approval	1992
Prescription Drug User Fee Act	1992
Dietary Supplement Health and Education Act (DSHEA)	1994
Food and Drug Administration Modernization Act (FDAMA)	1997
Pediatric Rule	1998
Medical Device User Fee and Modernization Act	2002
Pediatric Research Equity Act	2003
Critical Path Initiative	2004

Source: www.fda.gov

In 1994, the Dietary Supplement Health Education Act, which is typically abbreviated to DSHEA, was signed into law; it has had significant impact on the supplement industry. Next, in 1997, FDAMA, or Food and Drug Administration Modernization Act, initiated significant changes in the operation of the FDA. In particular, additional monitoring for what are called post-marketing studies was authorized. This act also codified several expedited review policies for certain medical devices.

The pediatric rule was issued in 1998 and several initiatives affecting development of pharmaceuticals and children have come into operation since; in particular, the Pediatric Research Equity Act (PREA) of 2003. Medical device user fees were introduced in 2002. Finally, the FDA identified a new strategic focus, the Critical Path Initiative, in 2004. The FDA, of course, hasn't been quiet since that time and, additionally, several important initiatives affecting clinical development have been implemented in the European Union (EU) and through the process of the International Council of Harmonization (ICH) as well.

In 2007, the Food and Drug Administration Amendments Act (FDAAA) of 2007 required all clinical trials (except the earliest studies in man — Phase 1 studies) to be registered in the clinical trial registry (ClinicalTrials.gov). In addition, the results of trials serving as bases for marketing approval were required to be posted.

1.2 Overview of Pharmaceutical Development Process

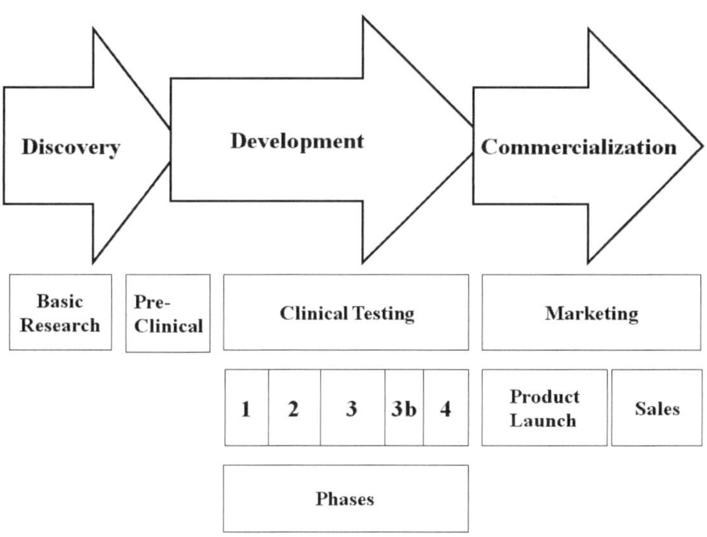

Figure 1-1. Pharmaceuticals Development Stages

Figure 1-1 describes, in a high level, summary fashion, the pharmaceutical product development process. The initial phase is discovery; at this stage, a compound is identified to move forward into development.

Next is non-clinical development – assessments in test tubes and animals. Then, the product's safety and effectiveness is assessed in people (clinical development). Finally, the product is delivered to patients (commercialization). In real life, there is a lot of overlap in these stages and processes.

We will be discussing these stages in much more detail in Part II of the book.

1.3 Contexts of Development

No development project is conducted in isolation. All elements of the environment in which the plan is conducted must be considered and addressed within the plan. There are several important contexts in which clinical development occurs. See Table 1-2.

Table 1-2. Contexts of Development

Contexts of Development
Health Care Context
Medical Context
Competitive Context
Regulatory Context
Time and Money
Development Process
Corporate Context

1.3.1 Health Care Context — The World in Which We Live

You don't have to delve too deeply into any newspaper today to understand that health care policy and health care costs have captured the attention of the public. Total health care expenditures for the year 2007 in the United States totaled $2.2 trillion. That number represents a 6.1% increase from the prior year and equals more than $7,000 per person in the US. That number also equals 16.2% of the US GDP. (Hartman, et al. 2009)

Prescription drug expenditures represent more than $200 billion of that number and increased 8.5% from the prior year. While there are several reasonable explanations for the annual increases, the matter of concern is that each year there are slightly different explanations for the increases, and each year the total health care expenditures increase. To keep yourself apprised of the latest figures, each year's January issue of the journal *Health Affairs* (www.healthaffairs.org) reviews the latest available data regarding health-related expenditures in the US.

Two other ways of considering the larger issues include the cost to develop a drug and total spending on research and development (R&D) by pharmaceutical companies.

Table 1-3. Estimated Cost to Develop a Drug (including failures)

Year Reported	Estimate
2001	$802 million
1987	$318 million
1975	$138 million
2006	$1,200 million (biologic)

Source: DiMasi J Health Economics. 2003;22:151-185

The estimated cost of developing a drug has skyrocketed to approximately $1 billion (a little less for a new chemical entity and a little more for a biologic). See Table 1-3. That seems like a staggering number — and it is — but it also includes the cost of failure. In large pharmaceutical companies some products succeed and become drugs, but many fail. Somehow, those failure costs have to be taken into consideration. The methodology of DiMasi and colleagues does this. Not all observers agree with the details of the numbers or even DiMasi's approach, but nevertheless, the numbers are large. Can a company, particularly a small company, develop a compound for less than this? Yes, no question. But that logic works only if the drug and the company are successful.

Another side of this coin is the total spend on R&D in the industry. Burrill and Company, on behalf of PhRMA [PhRMA, 2009], have estimated that all pharmaceutical companies together spent $65.2 billion on R&D in 2008. The R&D spend in US companies represented 20.3% of sales — a very large proportion. The question, of course, is what do they (the companies) and we (the consumers) get for all that money?

Indeed, the industry has a Problem (with a capital "P")! Over the past decade, US pharmaceutical expenditures have increased more than 2.5 times. The NIH budget has doubled. But, the measure of success, the measure of productivity — new drugs being delivered to the market — has decreased by half on an annual basis. Indeed, the trend has been fairly consistently downward over that decade. (See Figure 1-2.)

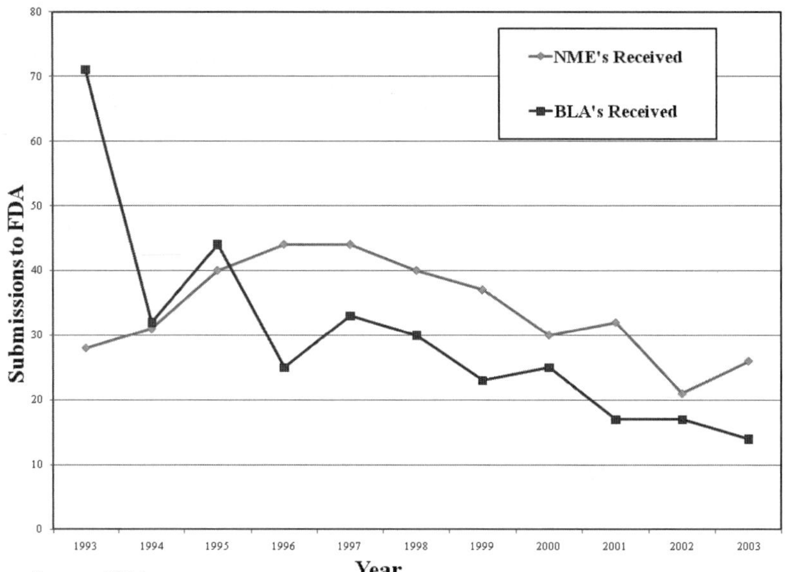

Source: FDA

Figure 1-2. Decreasing Submissions to FDA

Another way of looking at the cost of delivering a new drug to market is to isolate the costs of the individual phases of development. While pre-clinical costs have remained relatively stable, essentially all other areas of development have witnessed significant increases in costs. (See Figure 1-3.)

Previously, we noted a decline in the number of new NDAs and BLAs submitted to the FDA. Figure 1-4 shows that over the past decade approvals of drugs have flattened.

As you consider these data, keep in mind the time required for development — returns on a dollar spent today are not realized for several years.

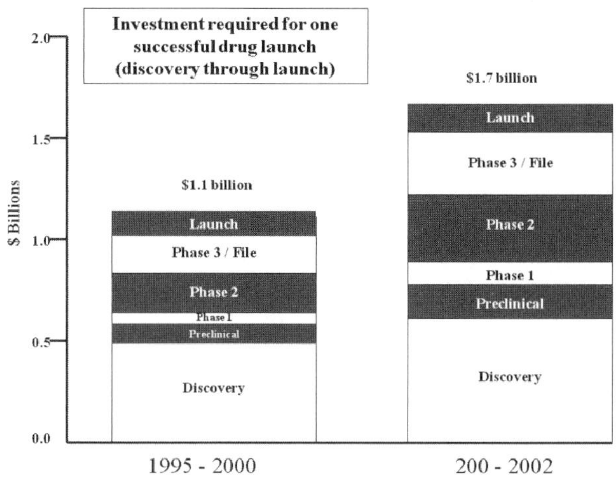

Source: Windhover's In Vivo: The Business & Medicine Report

Figure 1-3. Increased Investment per Development Stage

Source: FDA

Figure 1-4. FDA Approval Rate, 1994-2008

Another dimension of this rather dismal story is that there continues to be a high failure rate during clinical development. For drugs entering development in the mid-1990s, 30–40% failed in Phase 1. Of those entering Phase 2, 60–70% failed Phase 2. Of those moving into Phase 3, 45–55% failed in that stage. Finally, perhaps surprisingly, 10–20% of those products completing Phase 3 that were submitted for marketing approval were rejected by the FDA for various reasons. This represents a cumulative failure rate of >95%! Given that 10,000-40,000 patient trials are not uncommon in Phase 3, these are very expensive failures indeed. So clinical trials are probably too big, definitely too costly, and fail too often. (Wood, 2006)

What to do? Several initiatives have been introduced over the past few years by several of the important stakeholders. The FDA has introduced the Critical Path Initiative, the NIH its Roadmap, and the EU has embarked on a profound, well-funded effort combining the efforts of the Union, the member states, pharmaceutical companies, academics and regulators. The initiative is called the Innovative Medicines Initiative and was funded, in June 2007, to the tune of 2 Billion Euros – half of that money is coming from the European Union and half from industry. We will come back to these efforts in Chapter 15.

1.3.2 Medical Context

1.3.2.1 Product Advantages

When considering the development of a new product, it is important to identify the advantages the new product will bring to medical practice and patient care.

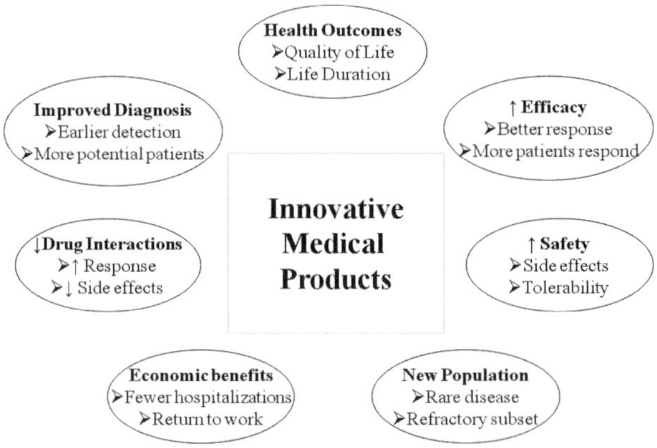

Figure 1-5. Potential Advantages of Innovative Medical Products

Figure 1-5 identifies some of the advantages a new product might bring. It could address improved efficacy, better safety, fewer drug interactions, better general health (thus providing economic benefits) and can address the treatment of populations of patients either not previously addressed or currently underserved.

1.3.2.2 Features of Target Indication

The next consideration within the medical context is the disease itself. We will come back to this topic in more detail when we discuss target product profiles (Chapter 3), but for now we can consider a big-picture view. (See Table 1-4.)

Will our product be applicable to all patients with the disease or only a subset with very specific characteristics? Does that translate to a lot of potential patients or relatively few? Is it so few that the disease might be considered an orphan disease? Is the therapeutic target concentrated in one or only a few areas of the world or is it a disease that knows no borders or boundaries? Is the target population concentrated in a specific group – older people, younger people, men, women?

As you consider entry of the product into the marketplace, you also need to consider whether it addresses a new indication for a previously approved drug – as such, you can build on a safety database as you explore the new indication. Perhaps it is an analogue of an approved drug; in this case, although the drug is new, experiences with its relatives might be very helpful in planning its development course. Or, perhaps, it is a new chemical entity (NCE) with which there is no prior experience.

As one considers development of a new chemical entity (NCE), several aspects are important to keep in mind. First, the compound will have an unknown safety profile and thus will require extensive chemical, physical, and biological characterization and quality documentation. The non-clinical (*in vitro* assays and animal studies) documentation will be extensive. Thus, an NCE requires acquisition of the most complete data set in order to be considered for marketing approval.

Another consideration is whether there is existing therapy for the indication or, perhaps, one is blazing a brand new development pathway. The latter approach, while exciting and potentially greatly rewarding, is fraught with a lot of challenges and risk.

An additional issue is the definition of the clinical target. Perhaps you are looking at hypertension – the definitions are clear and the measurements are very straightforward. On the other hand, perhaps you are considering a "new" disease. If the plan is to seek approval for a "new" disease, i.e., one newly identified or characterized in the literature, there are several risks associated with the plan. First, extensive documentation of the characteristics,

Table 1-4. Features of Target Indication

Breadth of Target

Can it be used in all patients with the targeted disease?

Is it useful for only a subset of patients with specific characteristics?

Extent of Disease

Are the lots of patients with the disease or only a few?

Are there concentrations of patients in specific geographic or demographic settings?

Individual Patients

All age groups or only pediatrics or only geriatrics?

Male only? Females only?

New Market Entry

Is it a new indication for a previously approved drug?

Is it an analogue of an approved drug (same class) in an existing indication

New Chemical Entity

Is this new chemical entity (NCE) to be used in an indication for which there is existing therapy?

Is this an NCE for a new indication (newly-defined or characterized disease)?

demographics, etiology, course (natural history) of the disease itself, and response to other therapies will need to be included in the submission – in addition to all the usual data referring to safety and effectiveness. Secondly, can a (reasonable) regulatory pathway be defined?

A good current example of a "new" disease is metabolic syndrome. It is a term that has come into much more common usage in the past few years. There have been several symposia and panels to look at, and define, criteria for the syndrome. A key observation has been the association of the syndrome with increased risk of cardiovascular events and death. But – and it is a significant "but" – not everyone agrees with the components of the definition. Furthermore, the association of the sequelae depends on the specific definition used. Worse yet, as of the present moment, there is no clear-cut regulatory pathway to approval – no drugs are on the market with metabolic syndrome as the indication and, even more problematic, the FDA isn't sure there really is such a thing as metabolic syndrome. Needless to say this therapeutic target is challenging at the least.

We noted above that one needed to consider the size of the population for whom the drug might be of benefit. If the population is small and fits into the definition of an orphan disease, there are some interesting and beneficial options available to the sponsor. We will discuss orphan disease in the section on Regulatory Strategy in Chapter 14.

1.3.2.3 Clinical Superiority

Finally, how will you establish clinical superiority for your product? There are at least two potential options – superior efficacy (it works better) and / or superior safety (it causes fewer side effects). Nirvana is demonstrating both! In any case, these two options become the primary focus of your clinical development efforts, but how do you measure them? Should the comparisons be with placebo or with some form of standard-of-care or some other agent? How much of an effect is required to be "superior"? Answering these questions at the earliest stages of development provides important guideposts for the whole development program.

Although we tend to think of clinical superiority in terms of better and safer, "better" might be convenience. For example, perhaps the new approach requires only a single injection per day rather than ten (OK, perhaps that is a bit obvious) or maybe it is trading intramuscular injections for subcutaneous injections (a bit more subtle). But convenience may be very important for patient management since increased convenience or comfort might improve patient compliance and that, in turn, improves disease management resulting in fewer sequelae or longer lifespans.

1.3.3 Competitive Context

You are almost never alone!

As you consider development of your product for its target indication(s), you must have an idea when your product will get to the market and then define and understand past, present and future therapies for the disease within the context of that time frame. Understanding these items will enable you to better position your product versus the competition.

In particular, you will need to consider what therapeutic approaches you will need to include in your development program. Some — those considered standard of care — might need to be included for all patients as baseline therapy. In other cases you might want to challenge the standard of care, but you need to be sure that patients are not put at risk in your studies. You will also need to consider what, if any, positive controls — currently marketed product, for example — you might want to include in late Phase 2 studies or the pivotal studies. (See Table 1-5 for a list of some key questions to consider.)

1.3.4 Regulatory Context

The next item to consider is the regulatory context. This is obviously an important component of any development program. We will introduce it here, address it intermittently throughout the book, and specifically address regulatory strategy in Chapter 13.

Table 1-5. Key Questions for the Competitive Context

What is the Current Management of Disease?
How good is it?
Are there cheap, off-label or generic options?
Is current management about to change?
What is on the Horizon?
Will management completely change?
Will products be additive or be substitutes?
What is in Development?
Who / what is your competition?
Are there new biomarkers that will make diagnosis / management easier or harder?
When Will You Get to Market?
What will be impact of new therapeutic approaches?
What will be impact of new products?

1.3.4.1 Goals of the Key Players

What is industry's goal in drug development? We need to get to market — otherwise we can't pay for the electricity and our children's school tuitions. We also want to get there as quickly as possible ahead of the competition, while using an optimal set of resources — time, money, and people.

What about regulatory agencies? They, naturally, come at the situation with a slightly different point of view. The US FDA defines its mission as "protecting public health by assuring the safety, efficacy, and security of human and veterinary drugs, biological products, medical devices", advancing the public health "by helping to speed innovations that make medicines and foods more effective, safer, and more affordable", and helping the public get "accurate, science-based information they need to use medicine . . . to improve their health." (http://www.fda.gov/AboutFDA/ReportsManualsForms/Reports/StrategicActionPlan/default.htm)

Even a superficial reading of the news suggests that many people think the FDA gets it wrong. However, the critics don't all agree about what is wrong — some believe the agency is too lenient, creating safety issues; others believe they are too rigid, preventing delivery of new products to the market.

The EMEA (the European Medicines Agency) describes its goals a bit more succinctly and seems to focus more on process and approvals. Its goals include providing a faster approval process, providing a better quality evaluation and achieving a lower global cost. (http://www.emea.europa.eu/mission.htm) However, we know they, likewise, care profoundly about the health status and safety of the populations of their member states.

1.3.4.2 Sources of Variability in Regulatory Processes

Although there has been progress since I entered the industry regarding guidances and guidelines and increased regulatory process transparency in many areas in the world, there is still a lot of variability built into the process. Some areas of variability are more controllable than others, but, in all cases, the sources of variability need to be identified and considered in the development planning process.

Table 1-6 identifies some of the sources of variability in the regulatory process in the United States alone. When combined with variability in processes in other countries and regions in the world, the whole process is complicated. I want to emphasize the last item in the list. Although identified as an issue for the US and the FDA, policy evolution is a fact of life in all regulatory agencies around the world. Even though regulations are written in black and white in all countries, there is a huge amount of grey in them. That grey enables a great deal of flexibility — some of it not so welcome — in the regulators' approach to sponsors. There are times (think early AIDS crisis) when the public clamored for faster, easier access to new therapies. Then, there are times (think Vioxx crisis) when the public clamored for tighter control and greater safety. In each case the FDA (in the case of the US) responded — in one situation, to facilitate development, and, in the other, by introducing additional hurdles.

Table 1-6. Sources of Variability in the US Regulatory Process

Different drugs and diseases
Different sponsors
Different FDA centers
Different FDA reviewing Divisions
Different reviewers
Evolving regulatory "climate" and shifting Agency policies

1.3.4.3 Factors Affecting Regulatory Process Evolution

As development program planning begins, a key issue is the time — 5 to 8 to 10 years — before the product gets to the marketplace. Since it is not possible to definitively know the regulatory environment that far in the future, planning must include contingencies to monitor the activities of regulatory agencies of interest.

Table 1-7 identifies some of the many factors that affect the evolution of this regulatory context. As noted, changing attitudes toward risk:benefit will create a pendulum effect. User fees have certainly facilitated some activities at the US FDA, but, approval rates haven't changed noticeably after the first year the fees were in effect. Yet, global harmonization has improved transparency and consistency across multiple markets.

Technology has been a bit of a plus and minus. There have been great leaps forward since I have been in the industry, but the technology also creates or (perhaps more appropriately) highlights some issues (e.g., electronic signatures). In addition, regulatory agencies have been slow to adapt new technologies, preferring to wait until more "standards" are in place. The recent focus on innovation and the "critical path" has generated a lot of excitement; positive results would be welcomed.

Finally, and not insignificantly, changes in world order play an impact. There has been a substantial movement toward conduct of clinical trials in non-traditional regions in the world, e.g., Eastern Europe, China, and India. There are opportunities and challenges here. India, for example, allows Phase 1 studies, but — currently at least — only for Indian pharmaceutical companies. Foreign pharmaceutical companies cannot test a dose level or dosing regimen in Indian patients that has not previously been tested outside India. That represents a significant challenge in the process of innovative development. Yet, the opportunity to extend trial populations to include patients in India (and other countries) facilitates patient enrollment and offers the chance for more rapid development. This is further balanced by the logistical challenges of conducting studies in a very large country with less than optimal roadways and communication.

1.3.4.4 The Global–Local Challenge

I have been doing worldwide development for more than 20 years. For me, the ICH process has significantly leveled the playing fields in Europe, the US and Japan. As a result, initiating and conducting worldwide development program is more feasible.

Table 1-7. Factors Affecting Regulatory Process Evolution

Factors
Changing attitudes toward risk:benefit
User fees (US)
Global harmonization (ICH)
Advances in technology • Information technology • Human genome • Standardization of processes
Focus on innovation and "critical path" issues
Changes in world order • Eastern Europe • China • India

However, the presence of ICH doesn't solve all of the local or regional problems. The ICH philosophy and many of the guidances and documents have been adopted by other countries that are not signatories to the ICH. However, when such guidances are grafted onto (as opposed to replacing) local and regional processes, attitudes, and regulations, there are opportunities for confusion and conflict. In any case, you must understand the requirements in the local environment in which you are working. The farther afield you get from traditional development areas, what you consider to be basic information may be harder to come by.

Finally, a crucial understanding is knowing that regulatory approval and product reimbursement are NOT the same thing in any country in the world. Therefore, all development programs must address reimbursement strategies for the markets into which one wants to sell the product.

1.3.4.5 Information Sources

There is a lot of (important) information out there.

For example, if one considers the United States alone, first there is the Food Drug & Cosmetic Act itself. Then, onto that is layered the Code of Federal Regulations (CFR) addressing preparation of INDs, NDAs, activities of Institutional Review Boards (IRBs) and labeling. In addition notices in the Federal Register are important. The FDA issues internal policies and procedures (MAPPs) and Guidance Documents (many of which incorporate ICH guidances). Then, there are advisory committees, policy statements, and policy initiatives — all of which affect the development process. The European Union also issues directives and guidances which affect development. Both regions issue summaries of approved products (in the US these are called Summary Bases of Approval (SBAs), in the EU they are called European Product Approval Reports (EPARs) that indicate the respective agencies' thinking about trial conduct and interpretation of safety and efficacy for various products and diseases.

Unfortunately, there is no easy way to stay on top of all the information, particularly what affects you. However, doing so is critical to your ultimate success.

1.3.5 Financial Context — Time and Money

Next, we move to time and money — the financial context.

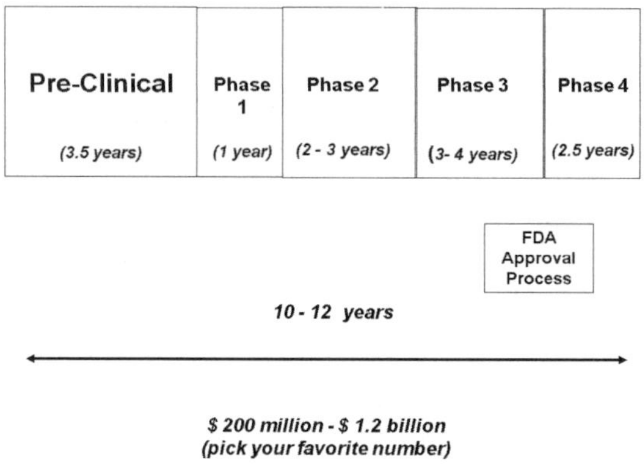

Figure 1-6. Time and Money for Product Development

Figure 1-6 depicts time and money in relation to the development process. The ranges are large and you can pick your favorite number as a hurdle. The importance of the relationship is not the specific numbers; rather, that developing any given product takes a lot of time and a lot of money and involves a lot of people.

Figure 1-7 describes in a very graphic way what happens in product development. If you'll look at the box on the left — Research and Discovery — it's a somewhat chaotic process. At some point, a candidate compound or compounds emerge(s) from here and moves on to the next stage. Graphically, the process is very much like a funnel, narrowing the candidate products, and — if we are lucky — delivering a product to the marketplace.

As we consider this process in the context of the company within which we work, we have to consider things such as the items mentioned at the bottom of the Figure — Pre-Clinical Compound Intensity, meaning how many compounds are in pre-clinical development, the number of compounds that fall out en route, the cost per compound — the ones that fail as well as the ones that are successful — and the development time per compound. Then, we need to calculate and understand the total development cost of each potentially successful new chemical entity. Ultimately, in a financial context, to be successful a company must use its capital efficiently.

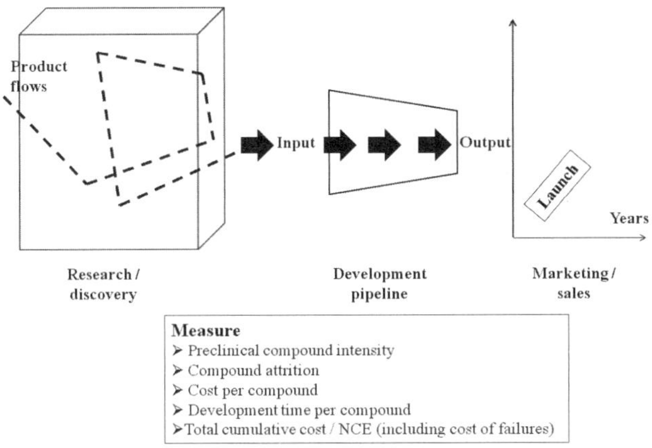

Source: McKinsey & Co.

Figure 1-7. Capital Efficiency in Research and Development

How is this efficiency achieved in practice? The next three figures look at this. PharmaProject and McKinsey surveyed a broad range of pharmaceutical companies examining decision making and its impact on profitability. On Figure 1-8, the attrition profiles of four companies are depicted. The companies — real companies — are designated as A, B, X, and Y. The Y axis represents the number of projects in the company, indexed to a value of 100; the X axis represents time in months. If you look at the profiles for each of these companies, they're different. The most remarkable thing, contrasting A and B on the top and X and Y on the bottom, is the period in time in which that 100 drops down to a significantly smaller number, say by, for example, 20 months. If you notice, for A and B, a significant number of compounds (projects) have been eliminated by that 20-month period of time. By contrast, for companies X and Y, the number of projects remaining at that time period is substantially higher. Remember — all those continuing projects consume some kind of resources — time, money, or people.

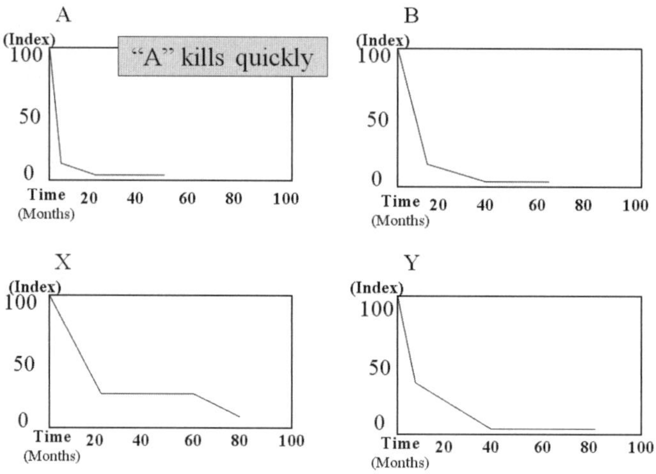

Index = Number of projects over time: 100 = total pre-clinical projects

Source: IMS R&D Focus; Pharma Project

Figure 1-8. Attrition Profiles in Product Development

Another way of considering the time-money question is to look at the same data but with an eye toward resource intensity — measured as spending per project month. In Figure 1-9, you will notice that there is a difference in our subject companies. Company A is highest on this scale, B is a bit lower and X and Y are grouped together at the bottom. So far A seems to kill projects faster and applies more money to the continuing projects per unit time than the other companies. The question is, do those behavior differences mean anything?

Finally, Figure 1-10 depicts several measures of productivity and profitability. Focusing for a minute on the right side, Company A seems to have higher productivity as measured by the number of new chemical entities (NCEs) delivered per unit of R&D spend and, in turn, as measured by sales per NCE launched.

Moving over to the left side of the graph, the payoff is represented as new sales per unit of R&D spending. A, and to a lesser extent B, are clearly differentiated from X and Y. So, the planning we do, the decisions we make, and our allocation of resources really do matter.

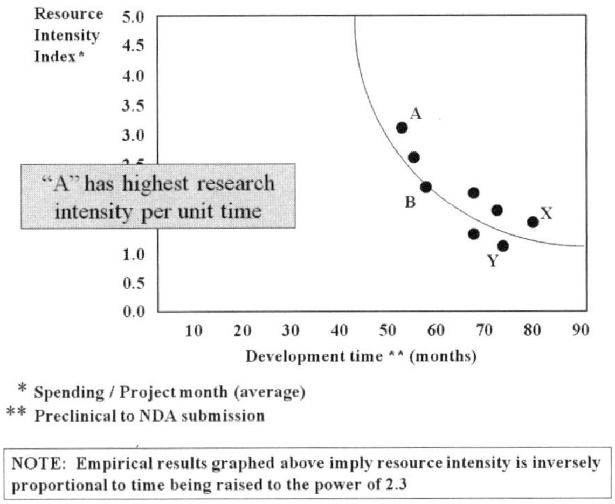

Source: McKinsey & Co.

Figure 1-9. Cost-Time Relationship in Development

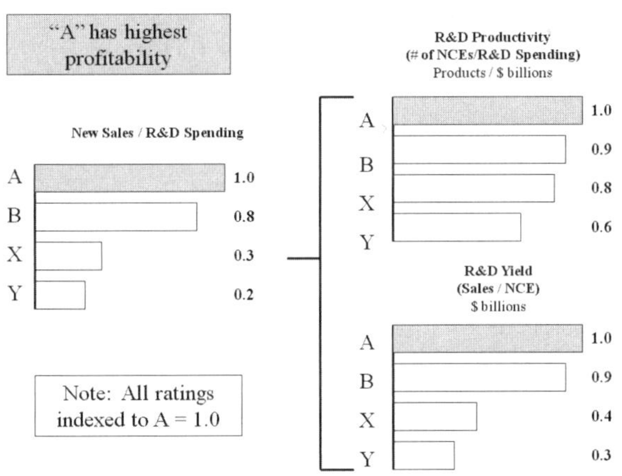

Source: Scrip League Tables; McKinsey & Co.

Figure 1-10. Comparative Research and Development Performance

1.3.6 Development Process

The next context to explore is the Development Process itself.

Previously, we cited the following odds of success: 1 in 10,000 from the point of discovery; of those compounds entering clinical trials, 70% get to the end of Phase 1, 35% to the end of Phase 2, 25% to the NDA stage and only 10% get to market. Given those odds, perhaps Las Vegas might offer more lucrative alternatives.

These numbers really represent the pre-combinatorial chemistry era. Nevertheless, they are illustrative. When I first joined the industry, it generally took a medicinal chemist about a week, on average, to come up with a new compound. Giving the chemist a break, we could expect about 50 new compounds per chemist per year. So, the numbers game in that environment was played by adding medicinal chemists. When you added all the numbers, only 1 out of 10,000 would ultimately make it to the marketplace.

Regardless of the initial number, the odds of success are pretty small.

However, there are other contributors to the story. The clinical phase of development, on average, is currently about 6 years. The approval phase is still about a year and a half, although it's become somewhat less over the years — particularly related to PDUFA funding.

However, within that average number are hidden two other factors — the therapeutic area of the proposed indication and the priority assigned by the FDA. Within categories of pharmaceuticals, there are significant differences; for example, neuropharmaceuticals take a significantly longer period of time (7.5 years in the clinic and 1.9 years in review) than do antibiotics — in particular, AIDS drugs (3.25 years in the clinic and 4 months in review). Other examples are endocrine and metabolic drugs (8 years in the clinic and 1 year for review) and respiratory drugs (5.1 years in the clinic and 2.25 years in review). (Mathieu, 2003)

Priority matters to the FDA and to other approval authorities. For products approved in 2004 by the FDA, the median approval time for all NDAs for priority drugs was 6.0 months and for non-priority (standard) drugs the approval time was 12.9 months. On the other hand, if only new chemical entities (NCEs) are considered, the median approval time was widely different. Priority drugs were approved in 6.0 months and standard drugs in 24.7 months.

So, for example, if a compound is assessed by those regulatory authorities as being a unique compound for a disease that doesn't have a lot of current therapy, then it's assigned a high priority. Then, a regulatory agency processes it in a quicker manner. On the other hand, if you're the fourteenth compound in this particular class for the same therapeutic endpoint, the priority goes down and your review time lengthens.

1.3.7 Corporate Context

Next, we will consider Corporate Context, that is, the strategic, tactical and stakeholder issues of the enterprise — the company — itself.

1.3.7.1 The Challenges

All companies face at least a core set of challenges. Costs rise. There are a lot of companies and products out there. If you are public or if you are VC-funded, your shareholders care a whole lot about how quickly you can get products to market. There is, as noted earlier, increasing scrutiny of the relationship between development costs and pricing. There are monetary constraints resulting in a limited access to funds; the changes in the world following the financial meltdown at the end of 2008 are illustrative of this issue. And, oh by the way, there are only 1440 minutes per day and that number never changes.

1.3.7.2 The Needs

Balancing those challenges is a series of needs: We really want it all, so we want the best products, the best allocation of development resources, the best product review and evaluation process (and shortest time, of course), the best markets, the best competitive profile and the longest time between introduction and patent expiration. That might seem a bit greedy, but in our heart-of-hearts, it is what we all really want.

1.3.7.3 The Environment

How do you work your way through these things? Well, first we have to start with understanding the business we are in and what our specific business objectives are. We also must specifically identify and consider the possible trade-offs we would be willing to make. Finally, the competition never goes away — we must know who / what they are and what are they doing.

1.3.7.4 The Big Picture

Let's focus a bit more on those business objectives. Specifically, are we interested in partnering? If so, at what stage of development? What about licensing? When would we do that? Or, what about taking it all the way to the end ourselves?

Those questions have to be put within the context of our current and projected cash position. How much do we have at present? What will this whole thing cost — including all operational costs? Thus, how long will the supply of cash last? Smaller companies, in particular, must consider that timetable and understand when they might need to raise additional cash and plan data generation around those needed fundings.

If we are headed to the goal line ourselves, we will have to address a set of commercial issues, such as obtaining regulatory licenses and achieving favorable reimbursement status in the target countries. We will also want to minimize the time to reaching peak sales while maximizing the overall peak sales. Unfortunately, there is nothing in this list that is a "slam dunk".

1.3.8 Corporate Context, Part 2 — Portfolio Management

Whether one is operating in a small company or in a large company, there are always limited resources — money, time, people — available to accomplish our goals. In addition, no matter how exciting a project might be at an intellectual level, it needs to be assessed — within the Corporate Context — in terms of its future value and its likelihood of success. Ultimately, you need to survive to be successful. Rigorous Portfolio Management is one tool for achieving that goal.

Portfolio Management is a process of managing multiple value-generating activities and assets within an organization to optimize future value. It means assessing, rating, valuing, and allocating resources in an "optimized" fashion to different technologies, different products (drugs, devices, biologics, combination products), different formulations, different indications, and different development strategies.

1.3.8.1 Objectives

The overall objectives of Portfolio Management include putting in place the best competitive quality portfolio possible, one that has both a high probability of regulatory approval and a high sales potential. The portfolio should optimally fit with the overall corporate strategy, taking into account core technical and operational strengths. Another objective is to develop a decision process within the company that enables the most-fully informed and best decision — resulting in operational consensus and cross-functional commitment.

1.3.8.2 The Process

To fully develop the subject of Portfolio Management is beyond the scope of this book. Nevertheless, I would argue the process needs to be a formal, structured, fact-based analysis. Even very small companies need to incorporate this discipline.

The key driver to analysis and decisions is future economic potential weighted by investment required. As we have noted, input on the probability of regulatory success and sales success — and the extent and quality of the competitive environment — need to be included and, in turn, all projects need to be prioritized. To frame the question: If there is a single resource and it can be used by either of two projects, which should get the resource? Allocate the resources according to that prioritization scheme. By the way, kill what should be dead. That is one of the hardest things to do in development. Remember Company A from our prior discussion.

Table 1-8 identifies a few key characteristics of a successful R&D organization. One of the key items on the list is to access necessary resources (i.e., consultants or other organizations) if not available within your organization. Also, realize that outsourcing is okay — for the right tasks to the right vendor. (We will discuss this more in Chapter 12.)

Table 1-8. Characteristics of successful R&D organizations

Clearly defines core competencies required for sustainable growth
Has access to group of world-class experts in each competency (inside or outside the organization)
Outsources noncore activities to partners or other experts
Includes specialists to manage vendor and partner relationships
Manages outputs and costs on a project-by-project basis

1.3.8.3 Key Concepts

This whole discussion sounds a lot like investing. It is. It is the same set of concepts taught in business schools when considering various financial instruments. There are some key concepts and terms that are useful to know within this context: opportunity cost, cost of capital, and call option strategy.

Opportunity cost is the value of an opportunity NOT chosen. By making one investment, you give up the opportunity to make another; the value of the second opportunity is the opportunity cost. One can put a value on that decision, where A is chosen over B.

Cost of capital really represents the time-value of money, i.e., a dollar in hand today is worth a dollar, but that same dollar in hand one year from now may not (in fact, most assuredly will not) be of the same value. We need to take that calculation into account. It is also the cost of using money. For example, we borrow money for a house or a company borrows money to build a factory. The cost of that borrowing is fairly easily determined — it is the interest / mortgage rate. Putting a value on the cost of capital is bit more challenging, but has historically run from 6-12%. To make an investment worthwhile, the investment needs to return a number (on an annualized basis) that is higher than the cost of capital.

One more concept: **call option strategy**. Taken to their ultimate, option strategies can become very complicated. For our purposes, I really want to keep it simple. When I purchase a call option, I have purchased the opportunity to obtain or purchase an asset in the future. A put option is the opposite. I have purchased the option to sell the asset.

In our context, I suggest thinking of the call option in the process of decision making, particularly in R&D. It will ultimately lead to better decision making. For a particular project, in order to decide whether or not to proceed, I would suggest defining the steps involved until the next decision point. Assess the value of proceeding versus allocating resources to a different activity. Once the decision is made to move forward with this project, you completely resource the project to the next decision point. The next decision point is, likewise, specified in terms of objective criteria. If the project continues to meet its objectives and continues to maintain its prioritization rank, the call option is purchased through each succeeding decision point. If the project doesn't meet the goals or it drops down in prioritization, different decisions are made — the option is not purchased.

1.3.8.4 Decision Analysis

The key to success in this approach is determining a *set of specific and measurable criteria* that are relevant to our ultimate objective. Ideally, they relate to the target product profile and / or to business objectives. The criteria would be mindful of the costs required to get to the next decision point (incremental costs) and the project's return. The *criteria must be established BEFORE results are obtained* and the decision point is reached (in order to avoid the inevitable emotional commitment that accrues to projects you "believe" in).

The review process must be disciplined and the decisions decisive. A major challenge in applying this model to the smaller company is the possibility that obtaining a "No" answer at a decision point, in effect, means discontinuing the only project the company has and, thus, closing the company. If that is the situation, that option (closure) should be seriously considered. However, if closure is a potential choice, I would suggest prior planning to formulate an alternative — redirection of the current project, a new project, merger, company sale or asset sale, etc.

As illustrated in Section 1.3.5 (above), companies with better portfolio management have better project selection, better project prioritization, better resource allocation and management, and are rewarded with appreciation in the value of the company and its assets.

References

DiMasi JA, Hansen RW, Grabowski HG. The Price of Innovation: New Estimates of Drug Development Costs. J Health Econ. (2003) 22:151-185.

Hartman M, Martin A, McDonnell P, Catlin A. National Health Spending in 2007. Health Affairs (2009) 28:246-261.

Mathieu, MP. *PAREXEL's Pharmaceutical R&D Statistical Sourcebook 2003/2004*. Waltham, MA: PAREXEL, 2003.

Pharmaceutical Manufacturers Association. *Pharmaceutical Industry Profile 2009*. Washington, DC: PhRMA, 2009.

Wood AJJ. A Proposal for Radical Changes in the Drug-Approval Process. New Engl J Med. (2006) 355:618-623.

Chapter 2
Overview of Product Development Process

Our task in this chapter is to cover the product development process — at least in an overview fashion. Putting it all in a single chapter reinforces the notion that development is a continuous process from beginning to end; understanding the end product (Beginning with the End in Mind) is crucial to laying the proper foundation on which to build the program.

Figure 2-1 summarizes the process from discovery through commercialization.

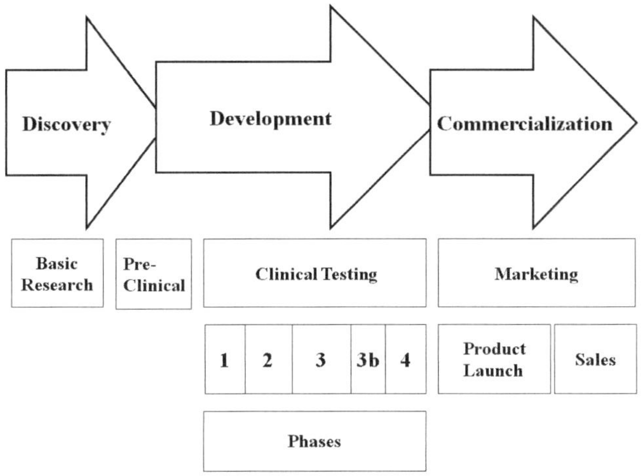

Figure 2-1. Pharmaceutical Development Process

As briefly discussed in Chapter 1, the four key stages are:

1. Discovery: In this stage, a compound or family of compounds is/are identified to move through development;
2. Non-clinical development: In this stage (depicted as a discrete "box" in Figure 2-1, but, in reality, continuing through the clinical phase), two key components are to:
   a) Develop enough safety and efficacy data to enable testing in humans; and
   b) Figure out how to make the material (drug substance and drug product) on a "sufficient" scale to enable testing;
3. Clinical development: During this phase, sequential observations are made to:
   a) Generate data enabling some understanding of how the drug is handled in humans;
   b) Permit some initial statements regarding safety;
   c) Determine responses (safety and efficacy) in a small group of patients with the targeted condition;
   d) Assess activity in broader and progressively larger groups of patients with the targeted condition; and
   e) Provide, finally, "substantial evidence" of the product's safety and efficacy in the group of patients to whom the product will eventually be marketed;
4. Post-approval development: At this point, additional clinical studies are conducted that might have been required as a condition of marketing approval or might be done to generate data for reimbursement, competitive positioning, or to extend the original indication to additional patient populations — in all cases broadening exposure and the safety experience.

Data generated during development ultimately need to fit into a submission for marketing approval. In the US this document is called a New Drug Application (NDA); in many other countries it is called a Marketing Approval Application (MAA). The format will generally conform to the elements described in Figure 2-2. The Marketing Approval Application is discussed in more detail in Chapter 14.

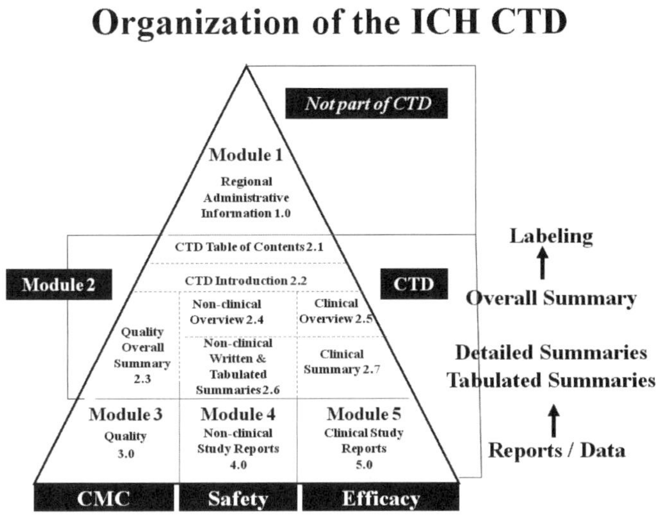

Figure 2.2. Organization of the Common Technical Document

Activities and processes concerned with Discovery are outside the scope of this book, so we will focus on Non-Clinical Development, Clinical Development and Post-Approval Development in this chapter. See Figure 2-3.

2.1 Non-Clinical Development

Reviewing pre-clinical development, we will focus less on the details of each particular component, but rather on the contribution each portion makes to the overall development process and, particularly, to clinical planning and execution. Our goal is to highlight those issues important to successful clinical development.

The three primary disciplines in non-clinical development are pharmacology, toxicology, and chemistry, manufacturing and controls (usually abbreviated CMC and dedicated to quality manufacturing of the drug). (See Table 2-1.) Two other topics of vital interest in non-clinical development are good manufacturing practice (applied to CMC) and good laboratory practice (applied to pharmacology and toxicology).

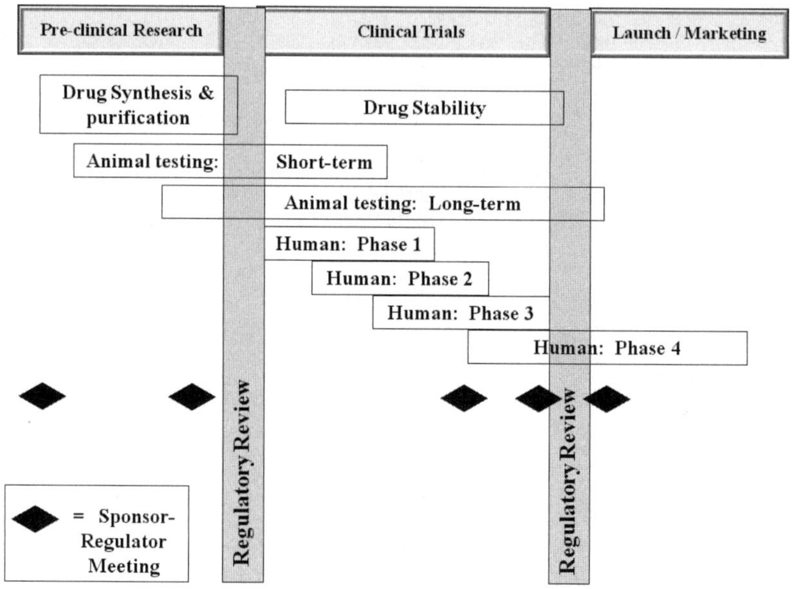

Figure 2-3. Summary of Development Process

Table 2-1. Major Components of Non-clinical Development

Chemistry, Manufacturing, Controls (CMC)
Drug Substance
Drug Product
Assay Development
Good Manufacturing Practices (GMPs)
Pharmacology — Toxicology
Pharmacology
Toxicology
Good Laboratory Practices (GLPs)

2.1.1 Planning and Execution

The name of the game throughout development is planning and execution. Although development is always, ultimately, dictated by the target indication and patient population and characteristics of the product, the sequence and type of studies in the pre-clinical portion of development are fairly uniform and — assuming good results — can be planned, scheduled and executed with reasonable certainty.

As a matter of practicality, you can't get to clinical trials without going through non-clinical development, so you might as well do the right things and do them the right way.

The goals of non-clinical development are to create and produce a reliable product, make educated guesses about the effect of the product in humans, and convince regulatory authorities that the proposed clinical trials may be conducted safely while providing data support for market approval.

The principles of success planning in non-clinical development are listed in Table 2-2.

Table 2-2. Planning Principles, Non-Clinical Development

Plan backwards, execute forwards
Let the overall development plan drive the non-clinical plan
Do not expose people to risks that have not been evaluated in non-clinical studies
Consult customers
Consult published guidelines
Concentrate on information flow

Plan backwards, execute forwards.

It is possible, at this stage, to generate a very long list of activities to perform during non-clinical development. However, some of the activities may not be relevant for the product you are trying to get to the marketplace or for the indication that you are pursuing. The task list should conform to the work needed — no more and no less. No more because you don't need to spend the extra resources and, clearly, no less since you won't get to the goal you want to achieve.

The first principle of planning for non-clinical development (and, indeed the whole of development) is "plan backwards and execute forwards". Begin this process by asking the following questions:

- Where are you going? (What is the target patient population and activity?)
- What do you want to have when you get there? (How do you convince a regulatory authority that your product is safe and effective? What are the data required to support that argument?)
- What do you want to accomplish? (Are you planning to take the product to the market or are you planning to partner prior to that point?)

As noted in Chapter 1, business goals and the ultimate product label must be kept firmly in mind when beginning the planning process and throughout the execution of the plan.

Let the overall development plan drive the non-clinical plan.

Begin this process on the clinical side of the equation. Rough out that clinical plan first. (See Chapter 3.) What is the population you would like to address? What is the anticipated overall duration of therapy — short-term (a course) or chronic (for months or years)? Competitively, how does this relate to other current (and potential) therapy? As development progresses, how will the duration of exposure expand during the clinical program?

What will be the doses and dose schedule and route of administration? Would loading doses be given or would an escalating dose be more appropriate? Will the drug be given alone or in combination with other therapy? Will it be given once a day, multiple times a day, or perhaps only once a week or once a month?

Are there differences in the geographic location of the population? Would a population handle this drug differently depending on where it comes from, from a geographic or ethnic point of view?

These are the same questions posed during the discussion of the Target Product Profile in Chapter 3. Understanding the answers also drives the foundational information generated in non-clinical development.

People can't be exposed to risks that haven't been evaluated in non-clinical studies.

One way to frame the process is to clearly state (and understand) that we cannot expose people to risks that have not been evaluated in some way in non-clinical studies. For the first studies, a direct link must be made between the information generated in non-clinical studies and the argument that the first protocol to be conducted in man will be safe for the subjects involved.

The formulation, the impurity profiles, the degradation, products, etc., all need to be understood as much as possible prior to the time that human studies are initiated.

The desired route of administration also needs to have been tested in the appropriate animal model(s). One specific challenge relates to inhalant types of products. The animal models for these are a bit difficult to perform. Yet, if that is the route of administration we are going to use in man, then we need to figure out how to do the pre-clinical studies to assure ourselves that when people are exposed to this formulation delivered in this particular way, that safety in relation to exposure is understood.

Obviously, in life and research, there is no such thing as a 100% guarantee. After all, we are making extrapolations from animal research to human beings. In that extrapolation, from experience, it is known that there is not a 1:1 correlation and there may be information that is just not known at the time. Mistakes are always possible, but the point is to be able to say, within the limits of what can reasonably be assumed, that there are enough data in the pre-clinical program to provide assurance, as much as possible, that this first protocol can be conducted in man safely. (Are there enough qualifiers?)

With each subsequent protocol conducted in man, we will be extending the exposure of the product into the human population. With each extension, there will be increased risk.

The goal is to have at least explored as many of the key questions as possible in the pre-clinical portion of development

Consult customers.

As the plan is assembled, our customers need to be consulted. The definition of customer in this context is a pretty broad one. It includes company personnel (e.g., executives and the development team), collaborators (and potential collaborators), scientific advisory boards, regulatory consultants, etc.

A potentially helpful set of consultants or advisors includes members of FDA advisory panels. For example, if the product is an anti-hypertensive and it's likely to go through the cardiovascular advisory panel for the FDA, then some potentially useful advisors are those current members of the cardiovascular advisory panel. The list of members of the cardiovascular advisory panel is published on the FDA website. (http://www.fda.gov/AdvisoryCommittees/default.htm)

Consider the list and identify individuals who are in their last year or two of membership on that particular panel. Then, depending on their background and their experience relevant to your particular product, hire them as consultants. They will not be sitting on the panel at the point in time when your product will be presented to the panel, so conflict of interest should not be an issue. What is exceedingly helpful is to understand the overall working process of the panel, the key kinds of questions the panel asks and particular concerns (e.g., efficacy measures, safety concerns) of the panel and the respective division of the FDA. The goal is *not* to obtain confidential information; rather, to understand the process and key issues so the development program can address them. In 2006, the FDA issued a set of guidances regarding conflicts of interest for members of advisory panels. Thus, while these individuals can provide very useful insights, it is important to not compromise their effectiveness as current panel members.

Various other contractors, as well as other regulatory authorities and regulatory agencies to which the marketing application will be submitted, likewise become valuable sources of information as the plans are assembled.

Consult published guidelines.

Regulatory authorities publish various guidance and guidelines. US guidelines, ICH guidelines, MHW or Japanese guidelines are available for review. The EU has many disease-targeted guidances. Consult as many of the potentially applicable guidances and guidelines as possible before you finalize your program.

Concentrate on information flow.

As you think about the process, it is important to concentrate on information flow. Understand and define the information required to submit a satisfactory IND (US) or IMPD (EU). In turn, that information is connected to the information required to submit an approvable application for marketing.

Look at the customer list: regulators, internal development groups, external collaborators, etc., and identify who needs information and what information is needed. Determine when it is needed. In what format is it best supplied? Is it best provided in tables? Is it best provided in text? Can it be provided electronically? Is it provided on paper?

The process is not simply box-checking. "I did that study. I did that study. I did that study. Now, it's time to move to the clinic." Importantly, the information generated from each of those studies contributes to the understanding of the product as product testing is advanced into man. Therefore, maximize the value of the information.

2.1.2 Chemistry, Manufacturing, Controls

All the data generated in pre-clinical development and in clinical development need to be based on a reliable and reproducible formulation. At the end of the development process, as regulatory agencies evaluate the data in support of the product's registration, they need to be able to answer these questions (in the affirmative): "Is the formulation used in the process of generating data at each stage consistent? Where changes have been made, are the appropriate bridging studies in place to ensure comparability?"

If there is some question that the formulation or product used changes during the course of development, then those data will be questioned. In fact, I am aware of a company that changed its manufacturing process in the middle of its Phase 3 program. At the time they submitted their data to the FDA, the FDA recognized that the company had not done studies which demonstrated that version A and version B were exactly the same product. As a result, the data were analyzed by the FDA as though version A and version B were two entirely different products. Needless to say, the company did not have data sufficient to achieve approval with either their version A or their version B. Given the need to repeat both pivotal studies (total cost in the neighborhood of $25 million), that was a very high price to pay for the error in judgment.

CMC is concerned with the **drug substance** (or active pharmaceutical ingredient, API) and the **drug product** (including formulation, i.e., capsule, tablet, injectable, etc.), and the assays involved for both substance and product. Pharmacology and toxicology are the other two major components. In each area, there is a necessity to employ "good practices" — good manufacturing practices and good laboratory practices.

With CMC, the focus is really to develop an understanding of the information needed to move from non-clinical development into clinical development. There must be coordination between those folks working on the product, those working on assays and those putting together the clinical development plan. At the point when the project moves into the clinic, a reproducible assay for the active ingredient must be in place. The product needs to have been stabilized.

During Phase 1 of the clinical development program, the product technically is "manufactured", but manufactured on an initial pilot scale which means small lots. Occasionally, it is possible to do one or two or three early Phase I clinical trials out of the same lot. Or, sometimes, the lots are so small that only a single study can be done.

It is important to ensure that the material provided to patients is equivalent to the material used in the non-clinical studies. While the process for manufacture is not fully validated at this point, it is reproducible within the release specifications.

The assay for the active substance, however, needs to be put into place. Although specifications can still be fairly broad, the assay itself should be pretty tight and reproducible. At this point, it is important to begin the stability program. Ideally, product in the stability program is always significantly in advance of that required for the clinical trial. For example, 30-day stability is available for dosing up to 28-30 days; 90-day stability is available for studies up to 3 months of dosing, etc.

As testing moves to Phase 2, the scale for manufacturing drug product moves into larger scale-up batches, typically still made in the pilot manufacturing facility. It is also reasonable at this stage to begin the process of method transfer to a larger-scale manufacturing facility and to move to process validation.

Assay specifications are generally close to final, although some assay development work is done to improve efficiency, shorten times, and decrease costs. From a stability standpoint, studies documenting shelf life, storage requirements and packaging requirements need to be completed.

Moving into Phase 3 means looking forward to registrational filings. The drug product needs to be in its final formulation. The process for the manufacture needs to be fully validated and three batches need to have been manufactured at product scale. Those three batches are going to be batches that will be audited by regulatory agencies. The manufacturing process itself has been validated and is auditable as well. The assay for the active ingredient needs to have its final specifications and be as tight as necessary. And, all methods for the assay will have been validated.

2.1.3 Good Manufacturing Practices (GMP)

A full understanding of Good Manufacturing Practices is essential; good reference material is the Quality series of guidances from ICH. (http://www.ich.org/cache/compo/276-254-1.html)

I believe a key mindset, as one considers GMP, whether pilot manufacturing or full-scale production, is that quality must be designed and built in, not tested in. If quality is simply defined by the point in time when products are tested for release, that approach is likely to be very costly — in terms of money and in terms of lack of marketing approval.

To extend the concept of quality as designed and built in, it is best to utilize systems approaches as the manufacturing processes are developed and designed. That implies a full understanding of the interrelationship between the objects, the materials, processes, procedures and the people involved in the manufacturing processes. By considering the elements as elements of a whole and considering the multiple interactions in the processes, engineering the process and testing as an integrated system, the chances of success are significantly higher.

While a regulatory agency might consider the folks in your company to be really nice people, the trust level of the agency doesn't extend too far. To use a time-honored and well-worn phrase, if it wasn't documented, it wasn't done. The phrase "trust me" does not work in this circumstance. GMP rises and falls on the ability to document all the activities that took place.

GMP requires an extensive process of training, auditing, monitoring and validation. If the organization has no prior experience with GMP, I believe it is important to get help. Identify someone or a group with expertise in GMP who can assist and guide the set up and operation of the process. An alternative, of course, is to contract out manufacturing — any of the steps from pilot batches to full-scale commercial manufacturing can be contracted. While those vendors should already be operating at GMP, ensure compliance with GMP through formal audit (even if this work must be contracted out). The key is to make sure this is all done early enough to ensure smooth delivery of product throughout the development process.

(However, if a sponsor feels compelled to build and operate their own manufacturing facilities, I would caution strongly against building them too soon. More than a few companies have confidently moved into site manufacturing only to have the clinical product fail in the clinical phase of development. That is a big ouch to the pocketbook.)

As an illustration of the need to adhere to GMP, consider the case of the United States against Barr Laboratories. (United States v Barr Laboratories, Inc, Civil Action No. 92–1744; http://www.gmp1st.com/barr2.htm) Barr was indicted for failure to adhere to GMP practices in manufacturing. By the time the FDA enforced a consent decree, collected a large fine and rejected the product, Barr had suffered significant losses in terms of its overall stock price. So it is costly. Although the front-end costs might seem high, trust me, it is reminiscent of the old Midas muffler commercials — "Pay me now or pay me later." The price tag for "pay me later" is much higher!

2.1.4 Pharmacology

Next up in pre-clinical development is pharmacology. Pharmacology comprises several specific disciplines — pharmacological screening, pharmacodynamics, safety pharmacology, and pharmacokinetics. (See Table 2-3.)

Table 2-3. Pharmacology

Pharmacology
Pharmacological screening
Receptor targets
Metabolic pathways
Pharmacodynamics / Safety pharmacology
Dose-response relationships
Mechanism of action of drug / product
Pharmacokinetics
Absorption
Distribution
Metabolism
Excretions

2.1.4.1 Pharmacologic Screening

Pharmacologic screening enables us to understand the characteristics of our product in a series of well-known assays. As a result of this testing, we are better able to develop a picture of potential therapeutic benefits. For example, you might run the compound against various receptor systems in certain animal models to try to understand effects on blood pressure, heart rate, gastrointestinal activity, neurological activity, etc.

2.1.4.2 Pharmacodynamics (Safety Pharmacology)

In addition, we then need to look at pharmacodynamics as it applies to safety pharmacology as well as in terms of its overall efficacy. We need to understand the dose we give, the pharmacologic response that occurs, and to understand the relationship between those two. That understanding is critical to our ability to project the pharmacodynamics data into the clinical situation. We are also expanding our understanding of the drug's mechanism of action which gives us more of a window into efficacy and potential safety issues.

2.1.4.3 Pharmacokinetics

Pharmacokinetics studies enable us to understand the absorption, distribution, metabolism and excretion of our particular product, as well as potential breakdown products of the primary ingredient. We get our first inkling of the relative contribution of the parent drug and it metabolites to its activity and its toxicity.

2.1.5 Toxicology

Toxicology is the third major component (along with CMC and pharmacology) of the pre-clinical development activity. In the toxicology area, there is a whole range of studies, as listed in Table 2-4.

Throughout the toxicology program, remember that the focus is on timing and the contribution of the information contributed to the overall program, in general, and to the clinical program, in particular.

Table 2-4. Toxicology Program

Toxicology Program
Acute toxicity studies
Subacute or subchronic toxicity studies
Chronic toxicity studies
Carcinogenicity studies
Special toxicity studies
Reproductive toxicity studies
Genotoxicity studies
Toxicokinetic studies

2.1.5.1 Acute Toxicity Studies

Acute toxicity studies are required prior to any Phase I study. These studies begin to give us insight into the organs and organ systems targeted by the drug. In dose ranging, we will explore the maximally-tolerated dose and to identify what the no-toxic-effect dose is. Regulatory agencies are a bit uncomfortable if both of those dose levels can't be identified in preparation for the first-in-man clinical trial protocol.

In the acute toxicity studies, typically two species are studied, one rodent, one non-rodent. At least three doses are explored (vehicle is included as a control). Frequently, those doses will be in log distribution, e.g., 1, 10, 100 mg / kg as the dose range. Initially, a 14-day observation period and gross necropsy are needed to document the overall effects.

2.1.5.2 Subacute Toxicity Studies

Subacute or subchronic toxicity studies are those studies that last up to three months, typically again done in two species, at least one non-rodent. The point of these studies is to define the pharmacodynamic response of the drug to understand dose-exposure and exposure-toxic effect relationships. Ultimately, those data are used to extrapolate dose ranges for human studies.

As a general approach, consider a lower dose that is likely to be a no-effect dose (it would be good if it still a multiple of the dose on a mg / kg basis (or on an exposure basis, calculated as area under the curve (AUC) for 24 hours, for example) to be used in the initial studies in man). Consider a series of escalating doses that would enable description of a toxicokinetic curve. Finally, the maximum dose should show toxicity; in a sense, it serves as an internal control as a way of confirming exposure.

2.1.5.3 Chronic Toxicity Studies

Chronic toxicity studies are those lasting 180 days or longer. The drugs that go through chronic toxicity testing are the drugs that are going to be used on a chronic basis. If use of the drug is expected to always be on a short-term basis (e.g., an antibiotic), chronic toxicology studies can generally be waived (with the consent of the regulatory agency).

The models for chronic toxicology are typically rat and dog. In general, it makes sense, from a financial standpoint, to delay the start of the chronic toxicity studies until there is some indication of drug benefit and drug safety. Otherwise, there is a significant risk of wasting money if the drug fails early.

2.1.5.4 Carcinogenicity Studies

Carcinogenicity studies are typically performed with drugs that are going to be used for periods of time longer than six months, drugs that would be used on a chronic basis.

Several categories of drugs and drug use will require carcinogenicity testing. For example, drugs included in this category would be non-steroidal anti-inflammatory (NSAIDs) for use in arthritis. Another group of drugs would be those used by patients on a chronic intermittent basis. For example, a group of patients might use the drug for 10 days this month and then 15 days next month and then, perhaps 5 the next. But, when mapped out over a period of time, there is repeated exposure to that particular product under normal use by humans.

Another category requiring carcinogenicity testing includes drugs which have a chemical structure that in the past has been associated with some sort of carcinogenicity or abnormality. In general, peptides and proteins are exempt from this type of testing since the generation of antibody responses prevents chronic administration in the animal models. Models tend to be mouse and rat and the studies are conducted over at least two years.

2.1.5.5 Special Toxicity Studies

Certain special toxicity studies also might need to be performed, depending on the particular development approach. For example, studies appropriate for particular formulations or routes of administration might include parental or local tolerance studies. *In vitro* hemolytic studies, photosynthetic or studies in a particular animal model which might be more specifically relevant to a human disease or to a specific human age might also be needed. This is an area in which to think expansively.

2.1.5.6 Reproductive Toxicity Studies

Reproductive studies are also required. Interestingly, the three major ICH regions (EU, US, Japan) haven't agreed on the details of timing for the respective segments. (See Table 2-6.) Other than identifying the three segments — cleverly named Segment I, Segment II, and Segment III — we won't spend more time on them here.

Segment I studies evaluate fertility and reproductive performance and are conducted in male and female rats. Segment II studies look at teratology and are conducted in rat and rabbit. Finally, Segment III studies examine perinatal and postnatal development and are conducts in the rat.

2.1.5.7 Genotoxicity Studies

Genotoxicity studies are required before moving forward in clinical development. Typically, prior to Phase 1 studies, an Ames assay, using *salmonella* to detect mutations in a target gene or an *in vitro* assay in mammalian cells looking at chromosomal damage (generally, a mouse lymphoma cell line), is conducted. Once Phase 1 studies are nearing completion, in preparation for entry into Phase 2, the micronucleus assay (conducted in rodent to detect chromosome damage to hematopoietic cells) will need to be completed.

2.1.5.8 Toxicokinetic Studies

Toxicokinetic studies relate drug exposure (systemic or tissue concentration) and dosing (dose administered) to particular toxicological side effects. It can be further categorized by duration of exposure.

This important information should generally be incorporated in essentially all toxicology studies. Data from these studies enable more precise scaling of the initial human doses, particularly in terms of expected or potential toxicity. In fact, this approach is continued when administering the drug to humans later in the development program.

2.1.5.9 Timing of Toxicity Studies for Conduct of Human Trials

Table 2-5 lists studies required for the earliest parts of the human clinical testing, prior to and into Phase 1. Also, the animal ADME (absorption / distribution / metabolism / excretion) data need to be obtained by the end of early Phase 1. They are not absolutely necessary to start Phase 1. However, many developers will complete them prior to Phase 1 to identify potential issues that can, in turn, be planned for.

Table 2-6 focuses on carcinogenicity and reproductive studies. As noted above, the carcinogenicity studies are not needed prior to clinical trials. They need to be included in marketing applications and are generally conducted at some later point during the development program, generally during Phase 3 (at a point when the efficacy and safety

Clinical Trials for Pharmaceuticals, ICH, 1997 profiles of the compound would seem to justify the expense). Occasionally, regulatory agencies will negotiate the timing of data from the final carcinogenicity, e.g., data become available after marketing approval. That's clearly dependent on certain circumstances and requires a certain comfort level for the agency.

Table 2-5. Toxicology Studies to Initiate Prior to Phase 1

Study	Timing
Safety pharmacology	Prior to Phase 1
Exposure data	Prior to Phase 1
Animal ADME	By end of early Phase 1
Single-dose toxicology	Prior to Phase 1
Repeated-dose toxicology	Equal to or longer than human exposure
Local tolerance	Prior to Phase 1
Genotoxicity	Part prior to Phase 1; Remainder prior to Phase 2

Source: M3, Non-Clinical Safety Studies for the Conduct of Human

Table 2-6. Carcinogenicity and Reproductive Studies

Study	Timing
Carcinogenicity	Conduct during late Phase 2 and during Phase 3; include in NDA
Reproductive: Male	US / EU: Complete prior to Phase 3 Japan: Segment I prior to Phase 1
Reproductive: Female	US: Birth control OK for Phase 1 / 2 Segment I + II before Phase 3 EU / Jap: Segment I+ II before Phase 1 All: Segment III for marketing approval

Source: M3, Non-Clinical Safety Studies for the Conduct of Human Clinical Trials for Pharmaceuticals, ICH, 1997

As noted previously, the reproductive toxicology isn't quite so straightforward. Key advice — check out the ICH documents before proceeding. (http://www.ich.org/cache/compo/502-272-1.html)

Tables 2-7 and 2-8 summarize the repeat-dose toxicity testing required to support clinical trials of various durations. The left-hand column identifies clinical exposure and the other three columns identify data needed to support that phase or registration activities by geographical region.

Table 2-7 focuses on clinical exposure up to 3 months. Table 2-8 describes testing for longer clinical exposure.

Table 2-7. Repeat-Dose Toxicity Testing for Clinical Exposure up to 3 Months

Clinical Exposure	Rodents	Non-Rodents	Support
Single dose	2 wk 4 wk	2 wk 4 wk	US: I, II, III; EU: I, II Jap: I, II, III
≤ 2 weeks	2 wk 4 wk 4 wk	2 wk 2 wk 4 wk	US: I, II, III; EU: I, II Jap: I, II, III EU: III
≤ 1 month	1 mo 3 mo	1 mo 3 mo	US & Jap: I, II, III EU: III
≤ 3 mo	3 mo 6 mo	3 mo 3 mo	US & Jap: I, II, III; EU: I, II EU: III

Source: M3, Non-Clinical Safety Studies for the Conduct of Human Clinical Trials for Pharmaceuticals, ICH, 1997

Table 2-8. Repeat-Dose Toxicity Testing for Longer Clinical Exposure and Marketing

Clinical Exposure	Rodents	Non-Rodents	Support
> 3 months	6 mo	Chronic	EU: III
≤ 6 months	6 mo	6 mo	US & Jap: I, II, III; EU: I, II
> 6 months	6 mo	Chronic	US & Jap: I, II, III
Marketing	6 mo	Chronic	Clinical Use > 1 month

Source: M3, Non-Clinical Safety Studies for the Conduct of Human Clinical Trials for Pharmaceuticals, ICH, 1997

2.1.6 Good Laboratory Practices

As good manufacturing practices relate to CMC, so good laboratory practices relate to pharmacology and toxicity studies. Again, the focus is understanding requirements in general and understanding the relationship to clinical development in particular, concentrating on information flow.

GLPs are applicable to all forms of non-clinical *in vivo* or *in vitro* lab studies, including: toxicology, pharmacology and biopharmaceutics, and apply to all products, including human and animal drugs, medical devices, biologics, and combination products.

GLP requirements, however, don't preclude the conduct of preliminary feasibility studies in non-GLP formats. GLPs are not applicable to initial pharmacological screening studies, initial metabolic studies and, potentially, initial pilot studies. These are the studies in which a set of key questions is being asked: "Does this drug have an effect on this particular model? Does it behave in a certain way? What kind of dose ranging can be done? What kind of absorption, etc.?"

These studies are built into the program in order to obtain a quick "Yes" or "No". If data are generated indicating an absence of the expected pharmacologic effect, there is no point in spending a whole lot of money to go any further; time to quit! On the other hand, if positive data are obtained, the appropriate GLP types of studies required to support initial human testing and marketing approval can be planned.

Similar to GMP regulations, GLP regulations apply to the organization, the personnel within the organization, the facilities, the equipment, testing all of that equipment and the facilities in operation. Ensure that testing protocols, appropriate records, reports, storage, security, signatures, approvals, etc. are in place.

The Standard Operating Procedures (SOPs) addressing GLPs must cover all aspects of the laboratory experiments. In addition, as with GMP regulation, extensive training is required to the SOPs. The SOPs will specify that protocols and reports be prepared for each study and will require identification of an adequately-qualified study director for each study. The SOPs will also specify record retention procedures. When reviewing the SOPs, it is necessary to look through them asking, "If I follow this SOP can I, in fact, do what it says I can do?"

I would suggest the SOPs corresponding to GLPs be specific enough to enable and ensure performance, but with non-essential elements unspecified. Then, non-critical processes could be altered without having to re-write the SOP or issue deviations.

As with GMPs, it might be necessary to hire external resources to implement GLPs in the organization. Other requirements that must be in place involve other environmental activities. So, for example, the Environmental Protection Agency (EPA), College of American Pathologists — laboratory assessments; Clinical Laboratory Improvement Amendments (CLIA) — laboratory assessments; etc.

The essential value of GLP processes and documentation is to assure the organization and regulatory agencies that high quality research has been conducted and that the data generated are reliable.

On the other hand, the "teeth" part of the GLP regulations is that, if there is evidence they haven't been followed, regulatory agencies might not permit initial human studies or approve subsequent marketing applications. It is possible that an agency might refuse to consider data from an entire study. For example, one of the primary toxicology studies might be disallowed and, if those data can't be included, then the clinical trial program cannot proceed. You have to pass Go without collecting $200 and repeat that study.

An agency might also refuse to consider all data from an individual testing facility due to GLP issues. If all the toxicology work was performed at that site, the marketing application might be rejected. Definitely a bad outcome!

Keep clearly in mind the reasons for embarking on this journey — delivering products to patients that will improve their health and their lives. Repeating the mantra, "Begin with the End in Mind", carefully plan and execute each step along the way.

2.2 Clinical Development

2.2.1 Introduction

Clinical Development is all about balancing risk and benefit:

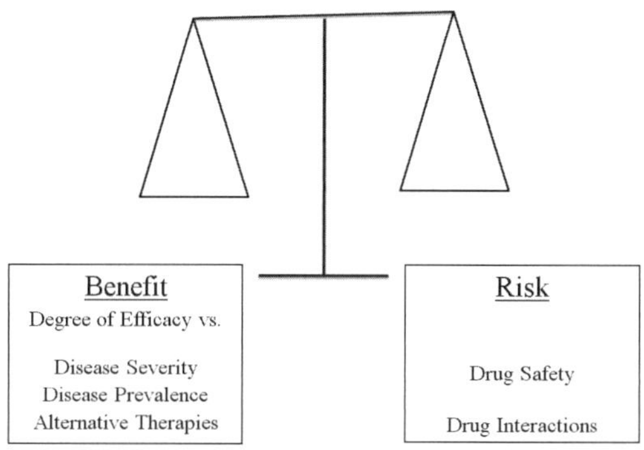

Figure 2-4. Balancing Risk and Benefit

In the Clinical Development Plan, there are several key milestones. See Table 2-9.

Table 2-9: Major Milestones in Clinical Development

Pre-IND Meeting
IND (includes Phase 1 protocol) Submission
Phase 1 Studies
Phase 2 Studies
End of Phase 2 Meeting
Phase 3 Studies
Pre-NDA Meeting
NDA (Marketing Approval) Submission
Phase 4 Studies

Consider that the work of the development program, but particularly the clinical development program, is to construct the appropriate risk: benefit profile for our product. Each study conducted and each analysis performed will enable us to add items to both sides of the scale. Hopefully, at the end, the benefit side is much heavier.

2.2.2 Clinical Development Logic

See Figure 2-1. Traditionally, clinical development is broken into phases, cleverly designated 1 through 4. The Phases have historically been referred to with Roman numerals, but recently there has been a transition to Arabic numerals.

There are fairly standard definitions of what constitutes each of the phases. In actual practice, there is a blurring of the definitions and, in fact, activities from different phases will be taking place at the same time. The traditional designations, however, are useful for discussion purposes. (See Section 2.6.)

Testing potential treatments in human volunteers to see whether they should be approved for wider use in the general population involves a series of experiments called **clinical trials** or clinical studies. The purpose of clinical trials is to demonstrate efficacy (clinical utility) in the disease process being studied while also providing evidence of the safety of the product in the study population. In order to proceed to clinical trials, the sponsor of the study (frequently a pharmaceutical or biotechnology company) needs to collect data *in vitro* and in various study animals (*in vivo*) in order to demonstrate that the first trials in people will be conducted with a reasonable benefit-to-risk ratio.

It is important to test pharmaceutical products in the targeted patient population. In the past, most new drug testing was been done on white men. Thus, groups such as women, blacks, and Hispanics often were inadequately represented. Since it is now known that drugs may work quite differently in people of various ages, races, ethnicity and gender, recently-designed study programs have been careful to incorporate, as much as possible, a wide variety of people.

Trial guidelines, including eligibility requirements, are developed by the study investigators and sponsors and usually include criteria for age, sex, type and stage of disease, previous treatment history, and other medical conditions. Clinical studies conducted early in a development program tend to include primarily healthy volunteers. Later studies will include individuals with the particular illness or condition to be studied. Within each study, the clinical trial protocol will describe which people might or might not be permitted to enter that study. This enables entry of the appropriate individuals for the evaluation of the drug, but excludes those for whom there would be unacceptable risk by participating in the trial. These criteria are detailed in the clinical study protocol. (See discussion of clinical trial protocol below, Section 5.3.)

2.2.3 But Who, in Their Right Mind, Would Do This?

The decision to take a new drug candidate into development entails a significant commitment in terms of money, resources, and time. The success rate, as we noted previously, for a product entering clinical trials to get to the market is, on average, approximately 10%. The development costs from discovery to market for a new chemical entity (NCE) range from $800 million to $1 billion by various estimates. More than two-thirds of those costs are consumed in the clinical trial process. The total time to develop a new chemical entity from discovery to market is, on average, more than 12 years.

Compounds (NCEs) fail at various stages in clinical development as noted in Chapter 1. There are various explanations for the failures and the explanations have evolved over the last decade.

In 1991, the main reasons for failure were:

Problems in Pharmacokinetics / bioavailability	40 percent
Lack of efficacy	30 percent
Issues with toxicology	12 percent.

In 2000, the main reasons for failure were:

Lack of efficacy	27 percent
Commercial and market reasons	21 percent
Issues with toxicology	20 percent.

Success rates also vary with therapeutic area. For example:

Cardiovascular	20 percent
Arthritis / pain	17 percent
Infectious disease drugs	16 percent
CNS diseases	8 percent
Oncology	5 percent
Women's health	4 percent.

Quite disturbingly, as noted in Chapter 1, the number of new chemical entities gaining market approval has decreased over the last decade to approximately 20 per year.

2.2.4 Preparation for Clinical Trials

In order to initiate clinical trials, a sponsor must first submit data showing that the drug is reasonably safe for use in humans in initial, small-scale clinical studies. During the pre-clinical stages of new product development, the sponsor must address several important issues in order to generate the data required for this submission.

First, does the product address the clinical problem? This is studied by evaluating results generated from *in vitro* assays to assess specific responses and *in vivo* animal models in which the pharmacologic response is studied. This enables the sponsor to understand the presence and the magnitude of the effect of the compound in models believed to have value in predicting the response in man. This includes evaluation of the absorption, distribution, excretion and metabolism of the compound.

Second, is the product safe? At this stage of development, this question is assessed in studies of *in vitro* and *in vivo* toxicity. The specific studies are, in general, prescribed by International Conference on Harmonization (ICH) guidances. At the preclinical stage, regulatory agencies will generally ask, at a minimum, that sponsors: (1) screen for genotoxicity; (2) determine the acute toxicity of the drug in at least two species of animals, and (3) conduct short-term toxicity studies ranging from two weeks to three months, depending on the proposed duration of use of the substance in the proposed clinical studies. (See Section 2.4.)

Third, can the product be manufactured consistently and in a form pure enough for additional studies? At the pre-clinical and early clinical stages of development, small quantities of the new chemical entity are required, but all components of the product that will eventually be tested in people must be understood from manufacturing and safety perspectives. The key is to provide the highest-quality compound that can be produced in a repeatable and consistent manner at a reasonable cost and in a timely manner. This will include documentation of the synthetic route and description and characterization of the active pharmaceutical ingredients (API) as well as inactive components and impurities. The Chemistry, Manufacturing and Controls (CMC) documentation will also identify process controls, product specifications, and evidence of stability.

Fourth, can a clinical trial protocol be developed that will permit evaluation of the product in people for the first time in a reasonable manner? A clinical protocol is developed that, in a rigorous and scientific manner, defines a question to be answered in the study of people (normal volunteers or patients with the disease to be studied) and frames the study in a way that optimizes the benefit of the study (information obtained, patient benefit) compared to the risk posed to the study subjects.

In the United States, pre-clinical meetings (Pre-IND meetings) are conducted with the appropriate review division of the FDA and are requested by the sponsor. Meetings at such an early stage in the process are useful opportunities for open discussion about testing phases, data requirements, and any scientific issues that may need to be resolved prior to IND submission.

In the US, the Investigational New Drug Application (IND) is a request for an exemption from the Federal Statute that prohibits an unapproved drug from being shipped in interstate commerce. Current federal law requires that a drug be the subject of an approved marketing application before it is transported or distributed across state lines. Therefore, in order to ship the drug across state lines, the sponsor must seek an exemption from that legal requirement. The IND is the means through which the sponsor technically obtains this exemption from the FDA. (See Chapter 14.) In the EU, the document submitted to obtain permission to initiate clinical studies is called an Investigational Medical Product Dossier (IMPD). It is typically a substantially shorter document than an IND, but summarizes essentially the same information. While, in the US, the IND is not formally "approved", the IMPD must be approved prior to trial initiation.

The regulatory agencies review the documentation regarding pharmacology (evidence of effect), safety (acceptable toxicity profile), CMC (appropriate characterization of the product and evidence of reproducible and acceptable manufacture), and the study plan (the clinical trial protocol). In order for the sponsor to proceed, the respective agency must conclude that the data provided are sufficient to indicate that it is, indeed, reasonable to proceed with certain human trials with the drug.

2.2.5 Clinical Trials

In order to obtain marketing authorization for a drug, the sponsor will submit compiled, analyzed, and summarized data detailing non-clinical (animal) and clinical (human) test data. The marketing application must provide sufficient information, data, and analyses to permit reviewers to reach several key decisions, including:

- Whether the drug is safe and effective for its proposed use(s), and whether the benefits of the drug outweigh its risks;
- Whether the drug's proposed labeling is appropriate, and, if not, what the drug's labeling should contain; and
- Whether the methods used in manufacturing the drug and the controls used to maintain the drug's quality are adequate to preserve the drug's identity, strength, quality, and purity.

Clinical trials represent the ultimate testing ground for unapproved drugs. During these trials, an investigational compound is administered to humans and is evaluated for its safety and effectiveness in treating, preventing, or diagnosing a specific disease or condition. The results of this testing will comprise the single most important factor in the approval or disapproval of a new drug.

2.2.6 Clinical Trial Protocol

Every clinical trial is (or should be) carefully designed to answer certain research questions. A trial plan called a **Protocol** maps out what study procedures will be done, by whom, and why. The Protocol is the heart of every clinical trial. It is a critical component of the planning, preparation, and execution of the trial. (If structured in a thoughtful way, it also forms the backbone of the final study report.)

The Protocol represents the planned course of action for the clinical study. It includes specific statements of the objectives and purpose of the study. It defines criteria for selection of the subjects to be included. The Protocol also specifies the design of the study. The design relates to the methods included in the conduct of the study to minimize the bias in obtaining (and then evaluating) the results of the investigation. It defines controls used as well as methods of blinding (i.e., preventing either subjects, investigators, other study personnel or all of the above from knowing which study treatment is used in which patient).

The Protocol also specifies the method(s) for determining the doses administered to the subjects as well as the observations and measurements taken during the study. Each procedure conducted during the study will also be identified and specified in terms of the timing utilized. Information will also be included regarding means of ensuring patient safety, particularly noting methods of notification in case adverse events are observed during the study.

Importantly, the protocol also identifies the specific methods of data analysis to be used to answer the research questions posed by the study. The format for clinical trial protocols is specified in ICH Guidances. The Clinical Trial Protocol will be addressed more completely in Chapters 5 and 6.

2.2.7 Development Phases

2.2.7.1 Phase 1

Phase 1 includes the initial introduction of an investigational new drug into humans. In general, Phase 1 studies address the following issues: estimation of initial safety and tolerability, determination of pharmacokinetics (PK), assessment of pharmacodynamics (PD), and potentially and early measurement or assessment of activity (efficacy).

These studies are closely monitored and may be conducted in patients, but are usually conducted in healthy volunteer subjects. (The exception to this is with oncology drugs. Because generally those compounds are expected to have certain toxicities associated with them, the initial studies — still called Phase 1 — are conducted in patients.) These studies are designed to determine the metabolic and pharmacologic actions of the drug in humans, the side effects associated with increasing doses, and, if possible, to gain early evidence on efficacy.

Phase 1 studies also evaluate drug metabolism, structure-activity relationships (SAR), and the mechanism of action (MOA) in humans. These studies also determine which investigational drugs are used as research tools to explore biological phenomena or disease processes.

The total number of subjects included in Phase 1 studies varies with the drug, but is generally in the range of 20 to 80, but might be larger if additional formulation and pharmacokinetic information is required.

The first Phase 1 study is usually a single-dose study where healthy volunteers receive a range of single doses of the investigational drug. The design and determination of the dose range relies on data such as the maximum tolerated dose (MTD) determined in preclinical animals studies. Vital signs and physiological parameters, such as blood chemistry, are closely monitored in the volunteers and the PK parameters in humans are determined for a single dose. The first dose utilized in these studies is usually a fraction (frequently 1/10th or 1/100th) of the no-observed effect level (NOEL) or dose in pertinent animal toxicity studies.

In Phase 1, you will be concerned about every single adverse event and every single lab abnormality. Your job at this stage is to put each of those observations into the context of drug exposure (blood levels) and into the context of your pre-clinical data. Understanding that information enables refinement of the development program. Occasionally, there are very early hints of potential problems that may affect the later elements of the development program. It's important to understand the potential problems at this stage so that future studies can be designed to generate data to clarify the issues.

The safety and PK data from the single-dose study serve as guide posts for a subsequent multiple-dose study in healthy volunteers, where indicated. Assuming PK, PD and safety are acceptable, one proceeds to multiple rising-dose studies extending observations to 10–14 days.

Examples of data collected in Phase 1 include ADME studies using radiolabeled tracers, studies evaluating the effects of food, and initial drug-drug interaction studies. For example, an antihypertensive drug would have a high likelihood of being considered for use in patients with concomitant congestive heart failure or renal insufficiency. Medications commonly used to treat those disorders, then, would likely be used concomitantly with the antihypertensive drug. Hence, it would be important to understand the drug-drug interactions (if any) prior to large scale trials and marketing. The key in Phase 1 is to get data quickly that enable decision making and better label writing.

Occasionally, Phase 1 testing is divided into two steps known as Phase 1a and Phase 1b. Phase 1a studies normally are conducted as a short-term study to ensure safety before embarking on a longer and more comprehensive Phase 1b study. Phase 1b studies can include actual patients and might provide first indications about drug efficacy against disease.

During Phase 1, sufficient information about the drug's pharmacokinetics and pharmacological effects should be obtained to permit the design of well-controlled, scientifically valid Phase 2 studies.

2.2.7.2 Phase 2

Phase 2 studies are the first studies in targeted disease states and tend to be larger (i.e., more patients, typically 100 to 300, and more centers per study) and longer (exposure measured in weeks rather than days) than Phase 1 studies. Nevertheless, trial design, through application of inclusion / exclusion criteria, in early Phase 2 restricts the population studied.

If the Phase 2 trials indicate the drug may be effective — and the risks are considered acceptable, given the observed efficacy and the severity of the disease — the drug moves to Phase 3.

Phase 2 studies are frequently broken into two sub-phases: Phase 2a and Phase 2b. Phase 2a studies are pilot trials designed to evaluate efficacy and safety in selected populations of patients who have the condition to be treated, diagnosed, or prevented. They may involve hospitalized patients or out-patients. Monitoring in this stage is still very close. Objectives may focus on dose-response, type of patient, frequency of dosing, or any of a number of other issues involved in safety and efficacy.

Phase 2b studies are well-controlled trials to evaluate safety and efficacy in patients who have the condition of interest. These trials usually represent a rigorous demonstration of a medicine's efficacy in a somewhat larger group of patients. In Phase 2, one is also beginning to explore pharmacokinetics, efficacy and safety in special populations, such as patients with the targeted disease and concomitant renal or liver impairment. Choices of these studies and populations are dependent upon the routes of metabolism and excretion of the drug.

Throughout Phase 2, key data relate to dose-range finding for therapeutic effect and safety tolerability. One aspect of this is correlating the pharmacokinetic information with the overall pharmacologic effect and with the adverse event profile. While each patient evaluated in the clinical program contributes to the overall safety profile, by the time Phase 2 data are available, enough patients have been evaluated that relatively common problems have been seen on a repetitive basis and can be put in perspective with regard to frequency and severity.

In my opinion, Phase 2 is where the game is won or lost. Companies frequently want to get through Phase 2 quickly so they can move to Phase 3, the pivotal studies. I would rather spend more time here, add a couple of studies if necessary, in order to better understand doses, dosing schedule, adverse event profiles, and efficacy endpoints.

While I fully understand issues with financial constraints and investor pressures, there have been well-publicized cases of companies that made mistakes by moving ahead with inadequate information. One company, in particular, decided they would sort out the doses in Phase 3. Unfortunately, data actually obtained weren't as clear as they might have been. Nevertheless, the company filed the NDA. The result was non-approval and the Agency essentially required a re-do of Phase 3. THAT was a very expensive mistake.

I absolutely believe that if you do not fully understand your targeted population, the doses and dosing schedule to be used, and the key measurable endpoints for efficacy at this stage, you should NOT proceed to Phase 3.

2.2.7.3 End-of-Phase 2 Meeting with FDA

In the United States, there is an opportunity to have a formal, face-to-face, meeting with the FDA near the end of the Phase 2 program.

There are several goals for the "End of Phase 2" Meeting. The first goal is to convince the Agency that it is safe to begin Phase 3 testing, the second is to review designs and endpoints for Phase 3 protocols. Finally, it is important to "preview" the information to be submitted in the marketing application. The preview is phrased along these lines: "These are the studies and this is the cumulative information we will have at the end of our Phase 3 program. IF (a big "if") the data are consistent with data collected so far and no additional issues are identified, will the package we are planning to submit be sufficient for review and approval?" The Agency, of course, can't guarantee an approval without seeing real data. The questions are geared at identifying any missing elements not currently in the development program AND, importantly, whether the Agency sees the current data differently than the sponsor. After reviewing the data presented, the Agency might have questions about the data that might only be answerable by generating more data. That enables the sponsor to collect the information and provide the "answer" in the original marketing submission.

One month prior to the "End of Phase 2" meeting, the sponsor will submit the background information and protocols for Phase 3 studies. This information should include data supporting the claim of the new drug product, chemistry data, animal data and proposed additional animal data, results of Phase 1 and 2 studies, statistical methods being used, specific protocols for Phase 3 studies, as well as a copy of the proposed labeling for a drug, if available. This summary provides the review team with information needed to prepare for a productive meeting.

A useful technique to focus the discussion around the Phase 3 program is to develop and share with the FDA a tentative draft of the proposed product label for the drug. If the sponsor has been using the Target Product Profile format in prior discussions with the Agency (see Chapter 3), details are being firmed up and requirements for Phase 3 data are relatively clear. Nail down the agreements with the reviewing Division at this point.

Again, the key objective of the end-of-Phase 2 meeting is for the sponsor and the FDA to reach agreement on plans for the conduct and design of Phase 3 trials. This, then, translates to agreements regarding the design and endpoints of the Phase 3 studies and that the studies (and the data package) will support the desired claims.

At the risk of bringing evangelistic tones to this discussion, let me reinforce the notion Do not go to Phase 3 without a clear plan (supported by data and approvable endpoints)!

2.2.7.4 Phase 3

The point of Phase 3 is to confirm therapeutic effectiveness. In this part of the development program, well-controlled, statistically-powered studies are conducted in multiple centers with adequate numbers of appropriate patients.

In Phase 3 trials, the drug is studied in several hundred to several thousand people with the disease. Efficacy and safety are further scrutinized. Efficacy is evaluated in the intended patient population, comparing the treatment to placebo or standard therapy (or both). Safety evaluation is also expanded; as more and more participants are tested over longer periods of time, the less common side effects are more likely to be revealed. Hence, the efficacy and safety profile of the product can be extrapolated to a larger, more general, intended patient population.

The larger controlled studies are generally referred to as **"pivotal" studies**. To be considered pivotal studies, they should be "adequate and well-controlled" (see below). The studies need controls (placebo, standard therapy, or both), have a design that is as unbiased as possible (generally randomized and with subjects and investigators blinded) and be of sufficient size to produce convincing results using appropriate statistical analyses. (See Chapter 6 for a discussion of trial design.)

The size of the studies (and, in turn, the size of the safety database at the time of submission for approval) depends on several factors: novelty of the product, availability of alternative therapies (and their relative safety), the size of the intended patient population, the intended therapeutic target, and the dose and duration of use of the product. ICH guidances suggest that, for chronically-administered drugs, the safety database include at least 1,500 patients with multiple-dose exposure (including 300-600 with 6-month exposure and at least 100 exposed for more than 12 months). In actual practice, programs tend to include many more patients.

A key balance in the Phase 3 program is enrolling enough patients of sufficient diversity to understand the safety profile of the product while maintaining enough homogeneity of the patient population to minimize variability in patient responses, hence enabling a reasonable statistical analysis of the data. Ultimately, the goal is understanding the benefit-risk relationship of the drug; that, in turn, provides the basis for writing the product labeling that will guide care giver usage following marketing approval. (Remember: The ultimate labeling approved by a regulatory agency generally will be limited to the patients [and their characteristics] actually studied in the development program.)

2.2.7.5 Pre-Approval Application Meeting with Regulators

The purpose of a Pre-Approval Application meeting is to discuss the presentation of data (both paper and electronic) in support of the application. The information provided at the meeting by the sponsor, within the context of the proposed product label, includes:

- A summary of studies (non-clinical and clinical) to be submitted in the Application;
- The proposed format for organizing the submission, including methods for presenting the data (within the Common Technical Document framework); and
- Other information that needs to be discussed (e.g., issues or problems).

In the pre-meeting submission package and in the formal presentation, data supporting safety and intended benefit in the intended disease target will be presented. As part of the presentation, key studies are highlighted including protocol design (e.g., endpoints, size, population, severity of disease), benefits demonstrated (e.g., "size" of effect, subgroups and differential responses, durability of response, consistency of data and analyses), and risks identified (e.g., overall safety profile, target organ safety, dose-toxicity relationships, data consistency).

It is also important to highlight ALL the issues (and their proposed resolution) as you see them. Since there is no such thing as a perfect drug (or a perfect submission), there will always be something to contemplate and to understand. Sometimes the issues involve manufacturing, sometimes toxicology, sometimes efficacy, and sometimes safety (or combinations). In any case, identify the issues up front. The regulatory agency may have other issues from their perspective, but the more you are "up front", the more cooperation you get from the regulators.

The meeting itself will focus on the development package (non-clinical studies, manufacturing information and clinical data), proposed data presentation formats, and the proposed labeling. It is critical that sponsors pay close attention to the discussions, resolutions of questions and the dynamics of the meeting to improve the odds of submission success.

2.2.7.6 Marketing Approval Application

Although the quantity of information and data submitted in marketing applications can vary significantly, the components are more uniform, particularly with the implementation of the Common Technical Document (CTD) format. There are numerous guidelines and guidances regarding content and format of marketing applications from both the FDA and the ICH. ICH guidances have generally been accepted by the FDA, but there are additional FDA guidances and there are therapeutic area guidances issued by EMEA that have not been formally adopted by the FDA.

The marketing application must provide all relevant data and information that the sponsor has collected during the product's research and development program. (See Chapter 14 for a discussion of marketing applications and the CTD.)

2.2.7.7 Advisory Committees

In the US, the FDA uses advisory committees to obtain outside advice and opinions from expert advisors so that final agency decisions will have the benefit of wider national expert input. We discussed this earlier in the chapter in considering potential advisors.

The FDA may especially want a committee's opinion about a new drug, a major indication for an already approved drug, or a special regulatory requirement being considered, such as a boxed warning in a drug's labeling. Committees may also advise the FDA on necessary labeling information, or help with guidelines for developing particular kinds of drugs. They may also consider questions such as whether a proposed study for an experimental drug should be conducted or whether the safety and effectiveness information submitted for a new drug are adequate for marketing approval.

Committee recommendations are not binding on the FDA, but the agency considers them carefully when deciding drug issues. (See Section 14.1.3.)

2.2.7.8. Post-Approval Trials (Phase 4)

Phase 4 trials are done after a drug has received a market approval. The trials may be initiated for several reasons. Such trials might be required by the US FDA as a condition of the approval of the drug for marketing. In this case, the Agency is particularly interested in longer-term side effect profiles or, potentially, activity of the drug in additional patient populations.

The sponsor may also initiate Phase 4 studies as a means of expanding the indications for which the drug is used, to gain additional data for marketing or to compare the drug to products currently on the market.

Phase 4 trials, particularly those required of sponsors as conditions of approval, have become increasing common. Indeed, recent data suggest that between 80 and 90% of recently-approved new molecular entities have Phase 4 study requirements included in the marketing approval communications.

2.3 Marketing

Once you get through all of this, and you are cleared for marketing, then what?

Before getting too excited about getting to market, here is a quick reminder. There are still some (actually a lot of) regulatory obligations that are consequent to marketing approval. The requirements vary a bit from region to region, so for products to be marketed multi-nationally, it is important to understand the requirements for each country and region.

The marketing plan created early in the product development process that has been modified many times subsequently now takes center stage and all that work and re-work finally pays off.

I suggest looking at the package insert as a component of your competitive marketing strategy. The label enables you to go into the marketplace and to sell the product versus other products already in the market. Thus, the desired labeling is the one most conducive to the most competitive profiling. HOWEVER, the marketing strategy is only as good as the data already in hand and already reflected in the product's label. The more competitive information collected during development (preferably in head-to-head studies), the more competitive the marketing approach can be.

Finally, the competition (products and companies) needs to be fully understood. Given the heat the industry has taken over the past several years for unethical practices, I might be a bit hesitant to emphasize this point, but I, nevertheless, believe the sales force needs to be thoroughly prepared, armed with data and aggressive, yet ethically based. Particularly for third-party payors (e.g., insurance companies, government aid programs), hospital formulary committees, national formularies and reimbursement authorities, well-thought-out presentations incorporating comparative data, particularly cost-effectiveness data, have the best chance of success.

Additional data might be forthcoming in the form of Phase 3B studies (begun while marketing applications were being considered) or Phase 4 studies (begun subsequent to approval). Typically, these studies address potential additional indications or expanding the labeling into additional patient populations. Other studies which might be included in this group would be pharmacoeconomics studies and quality-of-life studies. These studies should be planned early in the development program — they should not be after-thoughts — and be as comprehensive and as meaningful (i.e., utilizing important, economically-quantifiable endpoints) as possible.

Assuming the data are positive, such data should also be available to the marketing group; additional regulatory filings can be completed to formally expand the product's label.

2.4 Begin with the End in Mind

Revisit Figure 2-2. The goal of a development program is an approvable marketing application. By understanding the components of the application and building the plan to ensure that all components are developed appropriately and completely, the drug is provided the opportunity to win or lose on its own merits.

Fully understand the target indication, the market, the competition, and the regulatory requirements. Using the draft label (Target Product Profile) as a guide, develop a complete plan — through marketing. Plan with "the End in Mind" and execute for success.

Chapter 3

The Plan of Product Development and the Target Product Profile

In the Introduction we replayed the conversation between Alice and the Cheshire Cat in *Alice in Wonderland*. As the Cheshire Cat responded, to get from here to there, you have to know where "there" is.

Defining what and where "there" is represents the single most important decision in planning product development. Once that is known and understood, all plans can be enabled with that objective in mind.

Three key aspects of an approach to development include:

- Direction (arriving at the correct "there"),
- Speed (arriving "there" on time), and
- Efficiency (arriving "there" having used the fewest resources — time, people, money — necessary).

All plans should be designed and implemented with a successful outcome in mind. Indeed, I believe it is important to set the plans in motion, as noted above, so that the product — the drug, device, or biologic — wins or loses on its own merits. It is a tragic waste of resources to have a product fail because the "humans" messed it up through poor planning, poor design, or poor execution. On the other hand, if the product doesn't work or has a safety issue, hopefully that information is identified in the shortest time and with the least expenditure of resources possible.

This chapter will address creation of the Development Plan. We focus on a putative product profile that defines the ultimate goal (an approvable product and product label) and facilitates definition of the pathway (all the studies and activities) that gets us to that goal.

3.1 "Begin with the End in Mind" — the Product Label (Package Insert)

How do we start? Beating the proverbial horse to death, "Begin with the end in mind." That is useful advice for any planning exercise, but since we are concerned with drugs, devices, and biologics, "End" in this context means the product label and product approval.

The product label is most generally understood in the context of a drug (or a biologic product), so from this point forward, the examples cited will be primarily those relating to drug development.

The terms "**product label**" and "**package insert**" are used synonymously. The consumer understands the package insert as that piece of paper accompanying a prescription medication. The piece of paper has text printed on both sides in seriously small print. If the consumer has exceptional eyesight or a magnifying glass handy, he or she could decipher information that represents the most important elements of that product's development program. Summarized on the piece of paper are all the key statements regarding safety, efficacy and usage of that product and, in all cases, that information was developed through the course of the drug's research and development program.

(For more readable forms of drug labels, the most recent version is available through the FDA web site, Drugs@FDA: http://www.accessdata.fda.gov/Scripts/cder/DrugsatFDA/)

The key categories of information described on the package insert are listed in Table 3-1.

Table 3-1. Key Categories in a Product Label

1. Indications and Usage
2. Dosage and Administrations
3. Dosage Forms and Strengths
4. Contraindications
5. Warnings and Precautions
6. Adverse Reactions
7. Drug Interactions
8. Use in Specific Populations
9. Drug Abuse / Drug Dependence
10. Overdosage
11. Description
12. Clinical Pharmacology
13. Non-Clinical Toxicology
14. Clinical Studies
15. References
16. How Supplied / Storage and Handling
17. Patient Counseling Information

3.2 From Product Label to Product Profile (and back)

In Europe, the similar information is called Summary of Product Characteristics. For the sake of clarity, we will continue to refer to the document as a Package Insert or Product Label.

The Package Insert is a summary of all of the essential scientific information for the safe, effective use of the product. It must be informative and accurate, it cannot be false, and it cannot be misleading. The elements particularly related to the clinical aspects of the profile are based on human clinical experience. The statements must be based on patient trials that generated sufficient data analyzed in an adequate fashion, enabling one to understand the product's optimal usage in the context of its benefit:risk profile. Logically, the label must avoid making (or implying) safety or effectiveness claims not supported by data.

If the end product is a product label, we can use each of the headings in the product label (Table 2-1) to map out the aspects of the development program that would serve to provide the text for that section. To start, we create a Product Profile.

Decide on language (based on data) preferred for the package insert. Essentially, "fill in the blanks" in each section to reflect the preferred product profile. Fully assembled, that becomes a prospective Product Profile. This Product Profile is essentially a roadmap for development. It represents the intellectual scaffold upon which all the studies can be planned. It also serves as the managerial mechanism for streamlining and shortening the time from inception of the idea — the earliest point of discovery — to the point in time where peak year sales are achieved.

(Eventually, an *annotated* product package insert will be prepared. For each statement, under each heading, there will be a listing of data to support the statement. This document will, in turn, serve as the starting point for discussions with regulatory agencies regarding final product labeling — that single sheet with the microscopic print contained in the medication packaging.)

There is an iterative process involved in moving from Product Profile to Product Label. A Desired Product Profile (or Target Product Profile) is created and used to guide development. Once data are received, a revised Profile is created that, in turn, guides the next stage of development. Ultimately, the process yields an approvable Product Label based on the collected data. It is also possible, of course, that collected data do not support even a minimally-acceptable profile; in that case the project is abandoned.

I find that many companies do not go through this process at the start of product development and, as a result, they get to certain decision points and find decisions are guided by the emotions of the moment or the money left in the company's checkbook, rather than by reasonable, objective criteria.

3.3 Clinical Development Plan

3.3.1 Overview

As a foundational point, we might first ask the question: "What is the purpose of clinical research?" Fundamentally, I believe it is to answer scientific questions related to a disease process, the outcomes related to the disease or treatments (or, potentially, prevention) of the disease. In the context of drug development, the purpose is the delivery of a new therapeutic modality to the marketplace so that we can improve the health of patients with a specific disease process.

The two key questions for us to answer along the way are: "Is it safe?" and "Is it effective?"

Later in this chapter, we will focus on development of a Target Product Profile (TPP). I believe a well-designed and well-executed Development Program will essentially "fill in the blanks" for the topics identified in the TPP.

The sponsor will need to decide on the indication(s), the patient population(s) for whom the treatment is designed, the geographic scope of development (worldwide or confined to a single country or region), and key product features (those things that can build on current practice or current therapy while allowing discrimination in the marketplace). It is vitally important that the plan contain discrete, definable go-no go decision points.

As general categorical questions to identify the key go-no go decision points, consider the following questions.

> What treatment effects need to be demonstrated? Are the effects curative for the disease or are the effects disease-modifying? What is the minimum threshold of efficacy for this disease, for this population? How do these treatment effects compare to current and future therapeutic competition?
>
> What is the required dose range and dose schedule? What is an acceptable efficacy profile versus the (expected) safety profile? What about dosage strengths? How big will the tablets,

pills or capsules be? Might it be necessary to split them to make them palatable? Should other technologies for delivery be explored — inhalation, fast dissolution products, etc.?

What special populations need to be considered in the development plan? The basic special populations must be considered: hepatically-impaired, renally-impaired, women of child-bearing potential, elderly, pediatrics. Are there other populations we should specifically study? Perhaps, it might be a subset population of the primary therapeutic target population.

What is a "sufficient" safety profile? What do we know about the class of drugs or potential safety issues in the target population? What is our threshold for adverse events — type or frequency? Are their drug-drug or drug-disease interactions that should be considered at the beginning of the program?

Any development program will need inputs from multiple members of a team. At least the following disciplines should be represented: pre-clinical pharmacology, pharmacokinetics, toxicology, human pharmacology, clinical research (medical and clinical operations), statistics, regulatory, safety, medicinal chemistry, formulation development, and manufacturing and marketing. The list doesn't have to be limited to this group, but should at least include them. Small companies are likely to require the participation of individuals with some of this expertise from outside the company.

To develop the required background information, external consultant experts (including clinical key opinion leaders) may need to be consulted. Even the largest pharmaceutical companies in the world will utilize such individuals.

It is important to consult the appropriate and pertinent clinical literature regarding the disease, current therapies, and putative endpoints.

A way of "cheating" (not really!) is to look for summaries of other development programs for currently-marketed drugs for the target indication. In the US, summaries are prepared by the FDA and are posted on the FDA website. These are generally still referred to by the designation of Summary Bases of Approval (SBAs). (Assuming they are posted — older drugs might have minimal information on the website). (www.accessdata.fda.gov/scripts/cder/drugsatfda/)

In Europe, EMEA prepared summaries are called European Product Approval Reports (EPARs). (www.emea.europa.eu/htms/human/epar/eparintro.htm) The European summaries tend to be more integrated, although less detailed, than the US summaries. In either case, review them for key approvable endpoints, trial designs, sizes of trials, types of statistical analyses, potential safety issues, etc.

After collecting and processing the input, put together a plan — I suggest using the Target Product Profile template (See below, Section 3.3.2.3.) — that includes the range of studies to conduct, the designs of the individual studies, the target patient population, and the key clinical endpoints (efficacy and safety parameters). For the latter identify whether they need to be specific outcomes or whether an intermediate — surrogate — endpoint might be used. (See discussion of surrogate endpoints in Section 6.3.1.2.)

The results of identifying (and then executing) these plans will be to determine the range of efficacy, the optimal doses and dosing schedule, the required duration of treatments, baseline conditions (severity of disease or concomitant medications) that may facilitate or complicate the therapy being studied, and, finally, the safety profile of the product. In combination with efficacy, the sponsor is then able to construct a well-informed benefit:risk profile.

3.3.2. Potential Product Profiles

3.3.2.1 Target, Upside and Minimum Product Profiles

The various contexts we addressed in Chapter 1, particularly the medical and regulatory, will provide the essential information from which to begin to build the Profile. The key is for you to understand the product, the therapeutic indication, and the markets into which you plan to sell that product as well as the regulatory strategies required to garner approval.

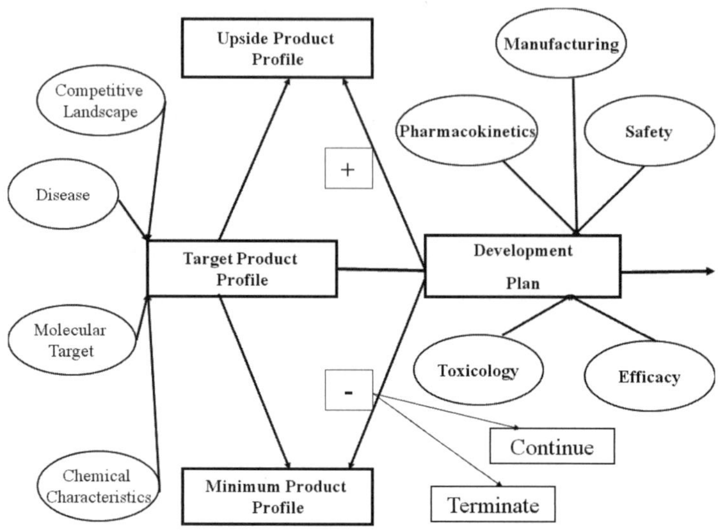

Figure 3-1. Product Profiles as Keys to Development

Figure 3-1 provides more detail about the elements that go into development of the various initial profiles and then the sources of data that will result in modification of the profile. Again, depending on the results obtained, we might continue on the path of development or discontinue if the data do not support even the minimal profile.

The first question is: "What's my ideal target profile?" So, if I had everything I could possibly imagine as a beneficial effect — efficacy, safety, convenience, drug interactions, product stability, etc., how would I translate that information to the categories of the product label we mentioned previously (Table 3-1)? Once the basic wording is agreed, this represents the Ideal Product Profile.

Next, allow a bit of realism to creep into the dreaming process. This is the somewhat painful acknowledgement that it is unlikely everything will be perfect. Now, the categories are filled in with statements that are more likely or probable. This is the Target Product Profile.

Finally, consider the "drop dead" scenario. This is the Minimum Product Profile. This is a crucial profile because it establishes the floor or threshold of product characteristics that are commercially viable. If the data generated during development fail to support that minimum threshold, then a "No go" decision should be made and either the project is modified substantially or dropped.

These profiles, as written down in the Development Plan, are used throughout the course of development. Hence, the Development Plan is really a living document. As newly acquired data become available, the Document Plan (and the prospective Product Profiles) is reviewed to evaluate the language of the projected profiles and the scope and details of the plans.

Each of the statements in the product profiles is tested against the newly-generated data. In view of that information, what remains and what must change? Next, the implications for the development plan are evaluated. Based on the data generated in this last study, is there a need to add a study or can one potentially be eliminated? Are additional non-clinical assessments needed? Is there a patient population that should be better understood in terms of this particular product and the therapeutic profile? Or, perhaps, do the data now suggest the patient population should be restricted for the use of this particular product? All of these observations have implications for the future of the project and our work plans.

The review of the data — vis-à-vis the product profiles — best includes all of the relevant stakeholders. I tend to regard "stakeholders" broadly and include management within the company, the clinical group, the regulatory group, and external advisors, as necessary given the size of the company, as well as the marketing group. At small companies, investors and the board of directors clearly fit into the definition of relevant stakeholders.

We briefly touched on marketing in the last chapter. In a former life, I was responsible for Phase 1 through Phase 4 for cardiovascular products. It was at that point that I fully realized the importance of having marketing people involved in the creation and shaping of the development plan at the earliest stages. It's a challenge, sometimes, for the marketing folks, because they look at the product and its stage of development and they'll say, "Gee, this product's not going to show up for another five years, seven years, ten years. Why should I put a whole lot of time into it today?" In addition to that, "You know, hey, there are going to be three or four more people involved in this project after me, so I don't need to get involved today." I think that's a crucial mistake at large pharmaceutical companies and I think it's a significant mistake at small companies. Marketing must be actively and repeatedly addressed in the development plan.

3.3.2.2 Constructing the Product Profiles

Refer again to Figure 3-1.

The clinical indications identified for the target profile represent the cornerstone of the whole development process. As such, be as specific and as aggressive as possible in the wording for the label. At the end, data will be required to support whatever text is in the final label, but it is good to include aggressive wording to make sure the appropriate studies, with the appropriate endpoints, are designed into the program. Consider whether all potentially-addressable indications will be the subject of the first regulatory submission or whether additional studies will be carried out in the peri- or post-submission period to expand the initial labeling.

As noted before, consider the competitive environment as the target profile is developed. As discussed in Chapter 1, considering the various contexts in which development is conducted, it is important to consider possible therapeutic approaches available at the time the product would be expected to get to the market, assessing the competitive space at the time of launch. Products available, or likely to be available, at that time must be considered and their relative strengths and weakness should be, as much as possible, understood. What is likely to be the "gold standard" of therapy at that time? Identify specific potential competitive opportunities and, in turn, identify the studies required to establish those features of the profile. Once this relative profile is understood, it can be determined if and when studies addressing the relative benefits / drawbacks should be conducted.

Start with efficacy. Define the critical outcome measure the product is expected to affect. Then, consider the degree of efficacy required as a minimum threshold for clinical significance and regulatory approval. How can the endpoints be measured and analyzed? For example, it is easier to quantify changes in cholesterol than it is to administer and interpret instruments assessing improvement in a patient with Alzheimer's. This part is crucial and requires an understanding of the current literature, approval pathways and labeling for products currently marketed for the disease, and current (and expected) regulatory guidances for the disease.

Then address safety. To determine the opportunities and requirements, the competitive marketplace must be understood — are there opportunities for the product based on therapeutic approaches currently in the marketplace? Understand what might be a no-go safety profile. Specifically, one might list any and all preliminary safety concerns (including limits of tolerability) for the active ingredient, formulation and route of administration and determine how the issues might be addressed in the course of development — whether through non-clinical or clinical studies (or both).

Drug interaction profiles represent an important subset of data regarding safety. It may not be possible to fully identify the interaction profile for the product, but as much information as possible should be included at the beginning of the program. Key considerations include drug-drug, drug-disease, drug-physiology (e.g., potential for abuse, tolerance, metabolic induction, etc.), drug-food, drug-gender, drug-race, etc. As additional information is obtained, additional drugs, conditions, or situations might need to be studied.

As noted above, actively create several profiles. The target plus optimal and minimal profiles will greatly help decision making as data begin to be generated. The Target Profile represents the most "realistic" one. The Upside Profile captures characteristics and attributes that might result in a more (or much more) favorable result. The Minimum

Profile is just that — the minimum acceptable for key parameters for either regulatory success or commercial success or both. Having this, as noted previously, facilitates the important go-no go decisions required during development.

The profile(s) initially developed, and the one(s) that evolve(s) in the process go generating new data, represent, of necessity, a series of compromises. Nevertheless, it is really important that the Development Team (and senior management) buy into the profile as created and own the process by which it will be updated and modified.

3.3.2.3 Target Product Profile and the FDA

The FDA, for several years, has had a pilot program (and website) for targeted product profiles. However, in March 2007, the Agency finally published a draft guidance incorporating their previous suggestions with input from various constituencies — including industry. Currently, use and submission of the TPP is voluntary. The draft guidance is located at http://www.fda.gov/downloads/Drugs/GuidanceComplianceRegulatoryInformation/Guidances/ucm080593.pdf.

The Target Product Profile (TPP) is consistent with the fundamental premise of this book — "Begin with the End in Mind". The FDA suggests that the purpose of the TPP is to "provide a format for discussions between a sponsor and the FDA" from pre-Investigational New Drug (pre-IND) phase through post-marketing regarding new indications (or changes in labeling for currently-approved indications). The format enables the sponsor to specify the desired labeling language for each item in the profile and the set of studies proposed to provide data supporting that language. The document would be useful at milestone meetings (pre-IND, Phase 2, pre-NDA) and would provide a check list for gathering data.

The FDA also envisions that the final version would be quite similar to the annotated draft labeling (mentioned above) that would be submitted with New Drug Applications (NDAs) or biologics license applications (BLAs).

Table 3-2. Key Sections of Target Product Profile

1. Indications and Usage
2. Dosage and Administration
3. Dosage Forms and Strengths
4. Contraindications
5. Warnings and Precautions
6. Adverse Reactions
7. Drug Interactions
8. Use in Specific Populations
9. Drug Abuse / Drug Dependence
10. Overdosage
11. Description
12. Clinical Pharmacology
13. Non-Clinical Toxicology
14. Clinical Studies
15. References
16. How Supplied / Storage and Handling
17. Patient Counseling Information

Source: FDA

The TPP is organized according to the key sections in the drug labeling and links drug development activities to specific concepts intended for inclusion in the drug labeling. (See Table 2-2 and note similarity to Table 2-1, Product Labeling.) Each section is further organized by Target (the wording the sponsor is hoping to achieve), Annotations (summary information of completed or planned studies), and Comments (providing additional clarity).

The sponsor drafts and updates pertinent sections of the template that are intended to support the specific statements in labeling. The sponsor then uses these updated versions of the TPP in preparation for discussions with FDA review staff to identify the most important development goals for the drug. The TPP is a dynamic ("living") summary that changes as knowledge of the drug increases.

Although the FDA's use of the TPP envisions a single Product Profile, I believe strongly that creating the Upside and the Minimum Profiles as well serves a useful guidance and discipline function during the process of Product Development.

3.4 The Development Plan

A Target Product Profile has been thoroughly discussed, vetted, and committed to a document. The appropriate upside and downside scenarios have been reviewed and included in that document. You have a roadmap for development. Now, it is now time to put together the rest of the Development Plan. See Table 3-4 for a suggested Development Plan outline.

In order to put this together, we must consider ALL of the elements regarding context that were discussed in Chapter 1. At this point, an outline of the information that is important to the organization is developed (see Table 2-3, as an example); from this outline each of the sections is completed.

3.4.1 Overall Success Potential

There are several key questions to pose to all the stakeholders (Table 3-3): What is the market potential for the target population? Are adequate resources available internally to conduct or lead all aspects of the development program? Are there any special issues regarding

Table 3-3. Key Consideration for Development Success Potential

Market potential — target population
Adequate resources to support development
Potential for manufacturing success
Potential for non-clinical success
Potential for clinical success
Potential for regulatory success
Potential for third-party reimbursement

manufacturing of the drug substance or the drug product? Are internal capabilities available to manage the scale up and manufacturing process? Are there any special issues regarding use of animal models for the product? Do we know that one (or another) species does not tolerate this class of drug and thus a different species should be used for non-clinical studies? Is that changeokaywith the regulatory agency?

Table 3-4. Development Plan Outline

	Executive Summary
1.0	Introduction
2.0	Market Opportunity
	SWOT Analysis
	Competitive Landscape
	Global Market Opportunity
	Life Cycle Management
3.0	Regulatory Strategy and Opportunities
	Target Product Profile (Summary) and Assumptions
	US
	EU
	Rest of World
	Post-Approval Label Expansion Strategy
4.0	Non-Clinical Development Program
5.0	Chemistry, Manufacturing, Controls
6.0	Clinical Development
7.0	Intellectual Property
8.0	Commercialization Strategy
	Reimbursement Strategy (by Market / by Indication)
9.0	Project Management
	Key Program Decision Points
10.0	Financial Summary (P&L)
11.0	Bibliography
Appendix 1	Projected Labeling / Target Product Profile (Detailed)
Appendix 2	Financial Model
Appendix 3	Development Plan Major Milestones
Appendix 4	Gantt Chart

3.4.2 Regulatory Strategy

Next, explore the scope of the potential regulatory filings. The strategy could include broad markets — North America, the EU, Eastern Europe, Nordic countries, Japan, Latin America, Australia, etc. — or could be more focused — single countries or single regions. This strategy must be developed in concert with the commercialization (and partnering) strategy).

As countries are identified and considered to be reasonable targets, understand the specific requirements for that country regarding labeling, approvals, documentation, and, importantly, reimbursement.

3.4.3 Non-Clinical Development Program

Assessment of the likelihood of success in the non-clinical development program is essentially an assessment of the risk:benefit ratio of moving the compound into clinical testing. Data from non-clinical pharmacology and toxicology studies provide answers to questions concerning potential efficacy in addressing the clinical target and the margin of safety in patients with the targeted clinical condition at the doses expected to be effective.

3.4.4 Chemistry, Manufacturing, Controls

What is the potential for manufacturing success? Have there been issues with manufacturing pilot batches for the initial non-clinical studies? Even if no issues have been identified to date, what might be encountered as scale-up is begun to prepare larger batches for non-clinical studies and, subsequently, batches for clinical trials?

3.4.5 Clinical Development

As discussed early in this chapter, there are lots of questions that need to be answered regarding the disease, current disease management, the endpoints of trials in that population, and any other special issues that will either facilitate or complicate development. What about regulatory success? We mentioned some of the key considerations regarding this earlier in this chapter. Finally, what about third-party reimbursement? If considering the US, is there already an appropriate CMS code for reimbursement? If not, how can an appropriate CMS code be developed?

3.4.6 Intellectual Property

Patent life is important. Since the patent clock starts early in the development process, the shorter the development time, the longer the patent life. The goal is to maximize the window between product launch and patent expiry. Early in development, consider strategies that will enhance the ultimate freedom to operate.

If a product is licensed in for development, do not skimp on the due diligence activities around patents, patent life, and freedom to operate.

3.4.7 Commercialization Strategy

The next part of the profile concerns the business strategy as it pertains to the development program. If the sponsor's plan is to out-license the product to a partner, at what stage of development will that happen? If out-licensing, co-development or co-marketing are potential strategies, what data (and in what time frame) will be required to "convince" the potential partner to "buy" our project?

3.4.8 Project Management

A project, by definition, has a beginning and an end. For development of pharmaceutical and biologic products and medical devices, the end is generally considered successful market approval (recognizing that our corporate strategy may involve partnering prior to that event). Project management involves the initiation, planning, execution, monitoring / controlling, and completion of the project. There are always constraints (people, time, money, quality, risk, etc.) that are managed and balanced in the process of reaching a successful conclusion. Given that in pharmaceutical development failure (drug doesn't work or isn't safe) can actually be success (go no further, re-allocate resources), the required go-no go criteria must be explicitly and straightforwardly defined. See Chapter 8 for additional discussion.

3.4.9 Financial Summary

The Financial Summary prepared for the Development Plan should reflect all operating costs — internal and external. One of the key operational questions has to do with the timing of hires versus continuing to use consultants. In some cases, it might make sense to use consultants for very long periods of time and, in other case, it might make sense to hire early. These decisions are largely determined by current core competencies, the complexity of development of the project and the presence (or absence) of other projects within the organization.

It is also important to project timing of approvals, market entry, and the revenue stream as it moves from initial revenues to peak revenues. (The latter might come in stages, depending on timing of entry into various markets and the competition in the respective markets.)

As part of this process, it is important to identify a target price. Although the market largely dictates prices, our estimate of the target price should incorporate cost of goods as well as an understanding of the therapeutic marketplace into which the product will be placed. It should also incorporate an understanding of the currently available regulatory schemes for the target disease in whatever countries are deemed to be targets for development and marketing.

Cost of goods should improve during the development process. However, estimates of unreasonable degrees of improvement and reductions in costs should be avoided at this stage. At an early stage, difficult and / or lengthy production processes may make the product non-viable commercially. Make a tentative decision about acceptable margins at an early stage.

In considering pricing, an important element is the concept of "added value". Essentially, it can be understood through the answers to these two questions: Why will the healthcare practitioner prescribe or use or recommend this therapy? Why will patients or insurers pay for it? This topic — added value — is a bit tricky. It is — at this stage — a bit of a wish list and is, in any case, very subjective. Nevertheless, I would suggest querying it to develop a better understanding of the marketplace.

Ultimately, one can calculate the return on investment (ROI) and answer the question, "Is it worthwhile to pursue development of this product?"

Chapter 4

Good Clinical Practice

4.1 What are Good Clinical Practices (GCPs)?

This definition is as good as any: Good Clinical Practices encompass the national and international regulations and industry-accepted standards that govern clinical trials on humans conducted to support applications to regulatory agencies for approval for marketing of drugs, biologics, and medical devices.

GCP Guidelines express the standards within the context of the design, conduct, recording of data, and reporting of results for clinical trials.

GCPs apply to everyone involved in the conduct of the clinical study. That includes: sponsors, monitors, clinical investigators (the physicians conducting the trial), contract research organizations that might be hired to support work in a clinical trial, as well as institutional review boards (IRBs) or institutional ethical committees (IECs) responsible for protecting human subjects.

In the course of this chapter, we will consider some of the background of GCPs and the principles involved. We will then discuss how the GCPs apply to investigators and sponsors. We will also describe the essential documents, defined in the GCPs, which are important for the conduct of clinical studies. Next, we will describe in some detail two of those documents — the Investigator's Brochure and the Clinical Trial

Protocol. One other aspect closely related to GCPs is IRBs / IECs; they will be considered in Chapter 7.

4.2 Background

Table 4-1. Good Clinical Practice Background

1947	Nuremburg Code: Rules for Conduct of Trials
1964	Declaration of Helsinki
1977	FDA Biomedical Research Program
1990	EU GCP Guidelines
1996	ICH GCP Guidance

Functionally, the process that has resulted in Good Clinical Practices began with the Nuremberg trials at the end of World War II. You might remember the Nuremberg trials dealt with, at least partially, some people who conducted medical experiments on prisoners without their consent. Out of the activities and discussions surrounding the trials came the Nuremberg Code which, for the first time, described a set of rules for conduct of research on human subjects. Next, in 1964, the World Medical Association compiled a document, entitled the Declaration of Helsinki, that expanded on the concepts of the Nuremburg Code. The Declaration of Helsinki has subsequently undergone several revisions. The current version is dated 2008 (Seoul). It is available on the World Medical Association website (www.wma.net/e/policy/b3.htm).

In 1977, the US FDA initiated its Biomedical Research Program as a way of monitoring the conduct of clinical trials. The EU first developed a single set of GCP guidelines in 1990. These served as the template for the ICH Guidelines on GCP issued in 1996.

The principles and the process of GCPs serve to provide assurance that the rights, the safety and the well-being of trial subjects, thus, anyone who would participate in a trial — whether it's a patient or a subject (otherwise known as a normal person) — are protected. By further defining a set of scientific and operational principles, conduct of trials according to GCPs, provide assurance that the clinical trial data are credible.

Currently, all regulatory authorities require an affirmation in applications for marketing approval that all clinical trial activities were conducted according to GCPs. Hence, GCPs affect all studies impacting the safety and well-being of human subjects and the generation of data for any regulatory marketing application.

Table 4-2 identifies those locations in the US Code of Federal Regulations, Section 21, dealing with Good Clinical Practices. Also listed is the ICH Guidance E6 which is the single most comprehensive document on GCPs and has been accepted by regulatory authorities worldwide as the "definition" of GCPs.

Table 4-2. Relevant GCP Regulations / Guidances

Regulation / Guidance	Topic
21 CFR, Part 11	Electronic Records, Electronic Signatures
21 CFR, Part 50	Informed Consent
21 CFR, Part 54	Financial Disclosure, Clinical Investigators
21 CFR, Part 56	IRBs
21 CFR, Part 312	IND Applications and "Responsibilities"
21 CFR, Part 314	Accelerated Approval
21 CFR, Part 601	Licensing
21 CFR, Part 812	IDE Applications and "Responsibilities"
21 CFR, Part 814	PMAs for Medical Devices
ICH Guidance E6	Good Clinical Practice

4.3 Principles

Clinical trials should be conducted in accordance with "the principles in the Declaration of Helsinki" and "all applicable regulatory documents". Those two phrases are incorporated into essentially all clinical trial protocols written today. Simple, but they encompass a great deal.

Table 4-3. GCP Principles

Conduct trial in accordance with principles of Declaration of Helsinki
Anticipated benefits of trial must outweigh risks
Available information must be adequate to support trial
Trial is scientifically sound
Medical care is responsibility of qualified physician / dentist
All individuals involved in trial are adequately trained
Subjects provide free and fully informed consent
Processes are in place to ensure adequate handling of the data
Data will be handled confidentially
Investigational products are produced in accordance with GMPs

The anticipated benefits must justify the risks posed to the individual subjects participating in the study. In addition, the rights of the subject, of the individual participating in the trial, prevail over the interest of science and society.

Also, the available information, both non-clinical data and other clinical data obtained in studies (and this is why we spent so much time in the last chapter going through the non-clinical information that's required to initiate clinical trials), must be adequate to support the trial in question. Hence, the proviso for "available information" includes not only the first clinical trial that would be the subject of the application for clinical trials, but all subsequent clinical trials as well.

In addition, the clinical trial must be scientifically sound. The point of this, of course, is that the "research" is not simply a fishing expedition. "Hey, let's see what happens if we use this drug (or if we use this device) for this particular person." Rather, there is a reasonable rationale attached to the clinical trial protocol. The endpoints of the protocol are well-defined. The statistical approach to the protocol is sound. Those elements contribute to the scientific soundness of the clinical trial. The clinical trial protocol is detailed; it's clear and has been approved by either an institutional review board (as designated in the United States) or an institutional ethical committee (IEC, the term used outside the US) PRIOR to the start of the study.

Furthermore, medical care is the responsibility of a qualified physician (it could be a dentist, depending on the particular type of study being conducted).

All individuals involved in the trial, that is, all persons who have to do with either the preparation of the protocol, training of the people who are going to actually implement the protocol, the people at the study site, the investigators, etc., must have adequate education, training and experience to conduct this type of a study.

Free and fully-informed consent is provided by the subject who is going to participate in the study.

In addition, processes and systems are in place not only at the investigative site, but also at the sponsor's location where the data are being handled, etc. to ensure that there is accurate recording of data collected, handling of the data, storage of the data and, ultimately, analysis of the data collected.

Further, the principles are crucially applied to confidentiality. We will talk a little bit later in Chapter 7 about what's called the Health Insurance Portability and Accountability Act (1996) or HIPPA. HIPPA has imposed new standards of confidentiality in all of medical practice in the United States and it does impact the conduct of clinical trials as well.

In addition, good clinical practices require that the investigational products themselves are manufactured, handled and stored in compliance with good manufacturing practices (GMPs).

GCPs also require having quality assurance systems in place. The point of these QA systems is that they are in place throughout the entire process, not simply for the trial under question. But, for example, for GMPs, for the manufacturing of the product, for the handling of data, for the training of individuals, etc., there are QA systems in place to protect the whole process.

To restate a couple of points we mentioned earlier, compliance with GCPs provides assurance that subjects' rights have been protected and the data are credible. As a result, compliance with GCPs facilitates the mutual acceptance of clinical data by regulatory authorities in Europe, the US, Japan and other countries and regions in the world.

4.4 Investigator's Roles and Responsibilities

The investigator's roles and responsibilities are outlined in Table 4-4.

Table 4-4. Investigator Roles and Responsibilities

Investigator Roles and Responsibilities
Qualifications and Agreements
Resources
Medical Care
Communications with IRB / IEC
Protocol Compliance
Investigational Products
Randomization Procedures
Informed Consent
Records and Reports
Progress Reports

4.4.1 Qualifications and Agreements

Investigators must be qualified by education, training and experience. The person needs to be a licensed medical practitioner as appropriate for the particular type of study. For example, for a study addressing congestive heart failure, typically the target investigator would be a cardiologist. As the investigator for this kind of a study, it might be even better to have somebody who is board certified as the clinical investigator, as opposed to, for example, somebody in a family medicine practice. However, there are many situations — for example in Phase 3 studies — in which targeted patients are those attending more general medical clinics, such as family medicine or internal medicine clinics. In that case, the family physician may be precisely qualified for the study in question. As the sponsor, the education, training and experience of the investigator must be matched to the protocol being conducted.

The investigator also must document what activities have been delegated; the investigator, however, needs to understand that delegation of activities does not mean delegation of responsibilities. He or she remains ultimately responsible at that investigative site.

The investigator must be familiar with the product. That information is primarily conveyed to the investigator via the Investigator's Brochure and from the protocol. In addition, the investigator needs to be fully cognizant of the design of the study, including all operational aspects. Furthermore, investigators need to have been trained to the principles of Good Clinical Practices.

The investigator must also agree to accept monitoring and auditing from the sponsor or from the delegated contract research organization. An acknowledgement of the acceptance is contained in the agreement the investigator signs at the very beginning of the study.

4.4.2 Resources

In the agreement the investigator signs, he or she acknowledges that they have an adequate number of subjects. This means subjects appropriate for the protocol according to the specified inclusion and exclusion criteria. Over and above the GCP issues, this is a practical matter. The sponsor wants to avoid the investigator who promises to enroll the desired number of patients but who, at the end of 6, 12, 18, or 24 months, will have enrolled only a single patient (and that patient didn't actually meet the inclusion / exclusion criteria). The sponsor invests a lot of money, a lot of time and other resources, ending up with little and, arguably, puts that single patient at risk — in the hands of an investigator who may not have been as experienced, or as interested, as advertised.

Another aspect of this is adequate time. This applies equally to investigators located in academic settings as well as investigators in private practice. They have other commitments and, if the investigator is in an academic center, they may be running a department, doing certain procedures, doing teaching, doing research, etc. The challenge is, "Do they have enough time on a weekly, daily basis to monitor and pay attention to the study?" If they don't — the sponsor really needs to spend time investigating this particular element with the investigator — then the site and the investigator should not be a part of this study. If there are questions about a site or an investigator, I would, in general, suggest trusting intuition — if there is a sense that the potential site or investigator is not going to contribute or have enough time, they probably won't.

In addition, investigators have to have adequate staff that is adequately informed. One of the challenges the sponsor frequently faces (I have heard this many times) is this comment from the investigator: "Ah yes, I just lost my clinical study coordinator about three weeks ago, but no problem, I will have a new one hired and trained by the time the study starts." That may well be true, but, on the other hand, if they have difficulty hiring or there are some challenges with the training, the sponsor may not be fully informed. Therefore, the sponsor should be cautious and not fully commit to that center until the new coordinator is in place and interviewed.

In addition, adequate facilities are required. Ask: "Do you have facilities to do the testing? Do you have facilities to store the binders for the documents? Do you have facilities for storing the investigative product, etc.?"

All of these items need to be fully evaluated by the sponsor's team before placing the study with an investigator.

4.4.3 Medical Care

The investigator needs to be a qualified physician who is responsible for all trial-related decisions in terms of the medical care of that particular patient.

In addition, the investigator is affirming that, "I, or my team, can provide adequate care for all potential adverse events." As a protocol is planned, the sponsor may well have some sense of what adverse events might occur because of prior studies (either non-clinical studies or previous clinical studies). On the other hand, there is always the distinct possibility that some unusual event might occur. In either case, the investigator must be prepared to deal with any adverse event. The agreement with the investigator states that the investigator and the facility have adequate capabilities and experience to be able to deal with an adverse event, should it occur.

I believe the primary physician, the physician who may have referred the patient for the study, needs to be informed of the information relating to the patient's participation in the study. This would include results of testing procedures and responses to therapy.

In a large academic center, a principal investigator is likely to have several sub-investigators participating in (i.e., conducting) the study. Nevertheless, the principal investigator's name is still on the FDA Form 1572 filed with the Food and Drug Administration (or the comparable form filed with other regulatory agencies). Thus, the PI is the responsible party. The investigator needs to understand what is going on in the study. Even if there are a number of sub-investigators participating in a study, documentation is needed that the primary investigator is fully informed and is in the loop of the communications. Physicians have been disbarred for failing to adequately monitor activities at their trial site.

The investigator is also responsible for communication to the sponsor in cases of patient withdrawal, adverse events, or operational issues that might arise during the study.

4.4.4 Communication with IRBs / IECs

Communication of investigators with IRBs and ethical committees is always a challenge. Frequently, investigators, particularly in major medical centers, may be part of an IRB. (That creates its own particular issues from the standpoint that the investigator cannot participate in discussions related to their protocol, other than answering questions, and cannot participate in voting relative to that particular protocol.) However, over and above those specific operational issues, investigators frequently aren't enamored of a review board out there watching over their shoulder; in turn, that investigator is less than excited about providing communication to the IRB.

Sometimes, the annual reports that are required by the IRBs are perceived to be onerous, annoying, aggravating, etc., and investigators don't like the idea of filling those out. However, it's absolutely crucial that the investigator establish communication with their review board. The investigator is also responsible for providing the appropriate documents, e.g., protocol, IB, consent forms, required summaries, etc. to the IRB.

With the initial review of the protocol, it is absolutely crucial that the first patient not be enrolled until there is a signed letter from the IRB stating that the trial is approved.

4.4.5 Protocol Compliance

The investigator agrees that he or she will comply with the protocol, that deviations require prior approval (except in emergencies), and that deviations must be documented.

This would seem relatively straightforward, but — let me tell you from lots of experience — protocol compliance issue is really a big deal.

There are physicians, particularly any of the members of the Dr. Famous clan out there at various medical institutions, who believe that the protocol is, in reality, only an approximation of how the study should be conducted. As a result, they might consider visits or procedures as negotiable items; they might consider restricted concomitant medication actually to be required. It can be sadly entertaining. Always do due diligence on your prospective investigators prior to involving them in the study.

4.4.6 Investigational Products

The investigator also acknowledges in the agreement he or she signs that they have facilities available and procedures in place to handle investigational products. That might mean, for example, locked cabinets, or it might mean temperature-controlled storage facilities, depending on the storage needs of the product. Sometimes those products need to be housed in a pharmacy instead of the investigator's facilities. The sponsor needs to work out all of these details and ensure that requirements are clearly identified at the beginning of the study.

The product needs to be accounted for throughout the study. A quick anecdote: At one point in my life, I worked for a company providing support for a pharmaceutical company that had engaged a certain investigative site. At this site, an investigator actually found a way to siphon off products and use them to treat various family members. Ultimately, when the monitor caught this, needless to say it turned into a significant deal. Eventually, the investigator was reported to the FDA and there were some significant ramifications for that particular study. Such an event is clearly out there on the fringe in terms of what happens during studies and at various investigative sites, but it really happened!

On the other hand, sometimes the people aren't paying attention. The sponsor will find files are missing. A specific package might be missing for a patient. It is absolutely crucial that there be adequate (i.e., complete) accountability for the all the drug product (and other trial supplies).

The investigative product needs to be used only in accordance with the protocol. The investigator can't decide to escalate the dose in a way that's different than specified in the protocol. They can't decide to alter the dose, give it at different times, etc.

Each of the products dispensed to the patient needs to be uniquely and adequately identified. If there are usage patterns for the product that are different than the typical once-a-day administration, the products need to be labeled and the patients appropriately educated.

Years ago, I was doing a study at a large pharmaceutical company and we did cross monitoring. The study was a large multicenter multinational study and one instance of the cross monitoring involved France and Germany. A monitor from France visited sites in Germany and someone from Germany visited sites in France. They monitored the first of what we called "trial patients" in a study. (We had decided that the first 10 patients would be evaluated in the study in this fashion and, then, depending on how they were handled operationally, we might have to modify the protocol.) Everything worked out fine except for one item:

The people in France and the people in Germany called Day One of the protocol a different day! One country labeled today Day One; the other country labeled today Day Zero and tomorrow as Day One. So, clearly, for us, it was really important to sort that out at the beginning of the study. That turned out to be a very valuable experience for us and clearly helped save this protocol. But, that's the kind of thing that you need to pay attention to. The patients in this case, across all centers, needed to receive directions to take the pill on Day One and all centers needed to count Day One in the same way.

4.4.7 Randomization Procedures

The investigator and all of his or her staff need to be fully educated and understand each of the randomization procedures involved in the study. Some studies have relatively complicated randomization procedures, e.g., calling a central randomization location to obtain the proper randomization sequence. Whatever the procedure is for the particular study, it needs to be well understood by all people who work at the investigative site.

Anything that results in premature unblinding needs to be explained. A frequent request to the sponsor is unblinding for an adverse event. The request, of course, needs to be put in the context of the study drug, the target indication, concomitant medications, and the event. However, I would suggest unblinding only as a last resort. Almost always, adverse events can be effectively managed without unblinding the study drug. In addition, by unblinding that patient, the person at the investigative site may detect a pattern. Ultimately, the investigator can figure out which is active study drug and which isn't. Although rarely a problem, it must be considered when dealing with these requests. Any premature unblinding for an individual patient or, obviously, for any group of patients, must be explained.

4.4.8 Informed Consent

The investigator also needs to fully understand the informed consent. In general, informed consents approved by the local IRB (or central IRB) will adhere to the basic elements of informed consent requirements.

There might, however, be local modifications to accommodate, for example liability procedures and site contacts regarding study information and medical questions. The investigator is agreeing that he or she is willing to fully explain the protocol and procedures, including risks and potential benefits, to the patient and is willing to answer any and all questions posed by the patient prior to study entry. This subject is important enough that we will return to it in Chapter 7.

The investigator also agrees that he or she will comply with regulatory requirements and adhere to ethical principles, that the consent will have been reviewed and approved by the IRB before use in the study, and that the consent will be updated, as appropriate, with new or additional information in a timely fashion. Furthermore, the investigator is agreeing there will be no coercion for participation or waiver of patient rights for any subject who will be participating in the study.

The investigator is also agreeing that the conduct of the trial will be adequately explained to the subject in adequate detail so the subject fully understands what will happen to them during the course of the study; when certain measurements or evaluations will be taken; when visits will be made; when they will take drugs or not; whether they have to use diaries or not. All of those aspects, the patient or the subject needs to understand; the investigator is agreeing that adequate communication will be provided.

The language that is used in the document, as well as the language that is used in the verbal explanation to the subject (or to the subject's representative), needs to be understandable. What is the primary language of the subject? If you are speaking to someone who understands relatively little English and you are attempting the explanation in English, that's a problem. On the other hand, if the person has maybe a grade school education and you try to explain this is very sophisticated medical terminology, the patient may again not understand. Either of those types of situations would not qualify as informed consent.

In addition, there has to be time for that subject to ask, and have answered, questions they have about the conduct of the study. What I can tell you from lots of experience is, there is no question that's too small. Sometimes the patient will have what appears to be, to you at least, a relatively trivial question. The answer may be relatively simple, but it's something that, if unanswered, might be a significant impediment to the patient's participation in the clinical trial.

Once the consent is signed and dated, the patient or the subject receives a copy and a copy is filed in the site records.

In the case of emergencies (this needs to be spelled out in the informed consent), what happens? Whom does the patient or their family contact? Details need to be clearly identified in the informed consent. By the way, the phone number or beeper number needs to work and, when used, put the subject in touch with a real, live person in a timely fashion.

Details of Informed Consent, including essential elements, will be covered in Chapter 7.

4.4.9 Records and Reports

In addition, the investigator must ensure the data are accurate, complete, legible, timely, and consistent with the source records. Verifying all of those things is dealt with in the monitoring process (discussed in Chapter 9).

The legibility piece is really, really important. Ask anybody who has ever had to enter data from case report forms (CRFs) into a database. Legibility is a huge issue. Many companies — Merck is a good example of this — have developed case report forms that require very little actual writing. They have put in as many check boxes as is possible for the investigator to use, so they don't have to worry about writing.

Completion of the CRFs needs to be done in a timely fashion. In your monitoring visits, you need to make sure that patients, even those seen in the past few days, have their data entered into the case report form. The only exception might be data that come from a laboratory; the data might not have been received by the time of the visit. However, the use of central laboratories with electronic reporting has largely eliminated this problem.

In addition, as part of the monitoring process, it is essential to ensure those data are consistent with the source records — whether it's an x-ray or a reported echocardiograph report, a laboratory report, etc. The data transcribed into the CRFs must be consistent with those source records.

Finally, the investigators also affirm they will provide adequate and appropriate progress reports to those individuals and groups that need them.

Records and reports also include any corrections that are made. For example, date of birth or age or yes or no or whatever. All of those corrections need to be dated and initialed and explained.

Financial arrangements need to be well documented. Currently, investigators need to disclose any potential conflicts of interest related to performance of the particular study.

Record retention is well defined in the regulations. Fundamentally, it amounts to two years after the last marketing approval in an ICH region.

4.4.10 Progress Reports

The investigator agrees to provide timely progress reports to the appropriate individuals at the appropriate time. The investigator acknowledges that he or she will provide, in an appropriate time frame, adverse event reports (details regarding the timing and conditions of reporting requirements are generally spelled out clearly in the protocol). Any protocol changes or amendments need to be communicated to IRBs.

4.4.11 Safety Reports

Safety reporting must happen. Absolutely has to happen. Any undocumented adverse events must be noted. The crucial question is what "undocumented" means. Fundamentally, if the adverse event is not documented in the Investigator's Brochure, meaning it has not been observed previously, then it's an unexpected adverse event. Then definitions of serious adverse events come into play. If you have an unexpected, i.e., "undocumented", serious adverse event, it needs to be reported to the sponsor and to the IRB. Other adverse events the investigator perceives are critical to safety must also be reported. For example, the investigator might note that in three or four or five of his or her patients, there are abnormalities in liver functions. If he or she perceives the events to be unusual for the investigative product or the disease, the investigator needs to contact the sponsor. Clearly, any deaths need to be reported as well.

Premature termination of the trial for whatever reason (financial, adverse event or other safety concerns) needs to be reported to the investigator's IRB or ethical committee. The investigator also commits that he or she will deal with the final reports necessary for the conduct of the study.

4.5 Sponsor's Roles and Responsibilities

Next, let's look at the sponsor's roles and responsibilities.

Table 4-5. Sponsor's Roles and Responsibilities

Allocation of Responsibilities
Trial Management
Monitoring and Safety
Investigational Products
Quality
Finances

4.5.1 Allocation of Responsibilities

Under GCPs, there is also an allocation of responsibilities to the sponsor. If the sponsor is a very large major pharmaceutical company, all of those responsibilities typically will be retained by the sponsor. On the other hand, if the sponsor is a very small pharmaceutical, biotech, or medical device company with a single regulatory person and a single clinical person who together are solely responsible for handling this trial, then there might be a need to assign responsibilities and allocate duties to other groups. This assignation of responsibility, if done, needs to be very clearly specified. In particular, it is identified in regulatory submission documents. Issues having to do with trial management, monitoring and safety, investigational products, quality and finances are responsibilities of the sponsor.

Groups of people, called contract research organizations (CROs), may assume the responsibilities for certain portions of the trial or for essentially all operational aspects of the trial. As we noted above, all assignations of responsibility by the sponsor to other individuals or organizations need to be reported to the FDA. As a basic rule, if it isn't specified, it stays with the sponsor. It is important to note, and we will discuss this in more detail in Chapter 13, that CROs are subject to the same regulatory review as the sponsor.

Essentially any activity, from earliest discovery through post-approval monitoring, can be contracted out to external vendors. Nevertheless, it never relieves the sponsor of the ultimate responsibility for those activities.

The sponsor is responsible for the ongoing safety evaluation of the product. While activities related to safety can be delegated, the overall responsibility for the safety evaluation will always remain with the sponsor. The CRO may run all aspects of the study including collecting, processing and analyzing, but in the final analysis, the sponsor is still responsible for that information.

The sponsor has an obligation to promptly notify all investigative sites regarding relevant and significant adverse information — the information might come from the current trial, it might derive from another trial being conducted with the same product, or it might even derive from non-clinical studies that are reported while the clinical trial is underway.

The sponsor is responsible for ensuring that each investigator participating in a multicenter trial will conduct the trial in compliance with the protocol. The investigator does not have the option to change it at his or her own whim. The sponsor also has the responsibility to put together case report forms to collect and capture the data at all sites.

Sometimes the simplest things can be extremely challenging. For example, if you are outside the United States, how do you interpret a date? In the US, the date 6-7-2008 is typically June 7th. On the other hand, if you are in France that date is actually July 6th. Those types of things need to be decided in your case report form and protocol; in any case, all sites need to adhere to the same process of entering dates. As a general rule, by the way, most studies in the United States now use dates in the European format. This means the number of the day first, the month written out or abbreviated as "Jul", for example, followed by the year, so the date above would be 06 JUL 2008.

It's also important to make sure that the same data are captured at all sites. If the study is conducted in multiple countries, the sponsor must decide about the language of the CRFs. Is it all going to be done in English? If so, then the sponsor must make sure the training for English is adequate in all those facilities. On the other hand, if it's collected in the native language, then the sponsor must make sure the data are handled in an appropriate fashion. For example, when I was at SmithKline, in order to address this challenge in a large multicenter, multinational trial, we made sure the blocks and the sizes of each of the rows on the CRFs were always all in the same location on the same page. Hence, if the text were translated from English to French or German or Italian or Mandarin or whatever, the space taken up for the explanation was exactly the same and the data entry boxes were located in exactly the same location from country to country. That process facilitated data entry and reduced errors.

In addition, the sponsor still has the responsibility for coordinating all participating investigators and to ensure they are clear about their responsibilities for the study. They have to make sure all investigators are similarly trained in the protocol. This is an example of the challenge: For a study, 10 centers have agreed to participate in the study. At the investigator's meeting, however, only eight sites have representatives. It is the responsibility of the sponsor to ensure that training for the other two was the same as for the first eight. As the trial moves forward, enrollment seems to stall and the sponsor decides to add an additional five centers. Likewise, the sponsor must ensure these five are trained similarly to the first 10. To cover those possibilities, many sponsors will videotape Investigators' Meetings as records of training.

4.5.2 Trial Management

The sponsor is responsible for all aspects of trial management. See Table 4-6.

Table 4-6. Sponsor Trial Management

Trial Design
Investigator Selection
Trial Management, Data Handling, Record Keeping, Independent Data Monitoring Committees
Submissions to Regulatory Authorities
Confirmation of Review by IRBs / IECs
Clinical Trial / Study Reports

4.5.2.1 Trial Design

The sponsor has the responsibility to ensure that qualified individuals are used throughout the study, including preparation and close-out. For trial design, these individuals might be in the sponsor's company, or they could be folks hired to conduct these studies. They need to be qualified in planning trials, putting together protocols, developing case report forms, planning for the analysis, doing the analysis, and writing the report. Individuals involved in the trial design process need to be people who understand how to do this, who are qualified, and who have the experience and the expertise to do it. See Chapter 6 for an in-depth discussion of trial design.

4.5.2.2 Investigator Selection

As noted previously, it is important to ensure that investigators selected for participation in the trial have adequate training, experience and resources.

When clinical trial processes are reviewed as a means of understanding how to improve conduct, study delay is typically an important issue. Frequently, the item associated with the delay is slow patient enrollment.

Poor enrollment may have any one of a number of explanations. One possibility is that the protocol wasn't designed adequately to capture the patients you really wanted to get into the study. Frequently, however, one of the key explanations is that the investigator(s) chosen didn't have the right patient population for participation in this particular study.

As part of the investigator selection, the sponsor should be able to provide for the investigator at least a close-to-final draft protocol and the Investigator's Brochure. Once the information is provided, it is important to review it with the investigator. Part of the review focuses on the practical aspects of the protocol. As written, can these procedures be performed at these time points, can this alteration in therapy be accomplished in the way outlined, etc.? The other part of the review focuses on the investigator's ability to identify and enroll the patients of interest for the protocol. That involves a careful review of the inclusion and exclusion criteria with the investigator and the investigator's staff.

Review of the Investigator's Brochure (IB) focuses on the mechanism of action of the product, past safety experience in non-clinical studies and clinical studies (if any), and other indications of clinical benefit. (The IB is discussed in greater detail in Chapter 5.)

It is the sponsor's responsibility to ensure the investigator's awareness of (or to educate the investigator and their staff about) all of the aspects of GCPs, including data handling, monitoring, reporting, etc. Site selection is also discussed in Chapter 9.

4.5.2.3 Operations, Data Records, Monitoring

The sponsor is, as noted above, required to ensure that all individuals involved in the study — at all levels — are appropriately trained and qualified. The sponsor is responsible for the data handling process and record keeping, even if these activities are contracted out to third parties. The data are the heart and soul of the trial; it is essential that they be handled correctly. In case a Data Monitoring Committee is required for the study, it is the sponsor's responsibility that it be composed of appropriately experienced individuals and that the charter and processes be in place to ensure adequate monitoring of the trial. Clinical Operations are discussed in greater detail in Chapter 9.

4.5.2.4 Regulatory Submissions, Ethical Review

Obviously, it is the sponsor's responsibility to provide all submissions to regulatory authorities. That responsibility is still in effect, even if CROs actually perform all the work. It is also important that clear and absolute confirmation be obtained that appropriate documentation is provided to the ethical bodies reviewing the trials and that information provided by the sponsor be the most up-to-date. As part of that responsibility, it is essential that IRB approval for each site be obtained PRIOR to the entry of any patient into a protocol-associated activity.

A hard lesson I learned years ago, when I tried to prepare my first NDA, is that all trials must be reported. That probably seems intuitive. However, for this NDA many trials had been completed (more than 100, actually) over many years and, of course, there were on-going trials that would be completed and would need to be reported. Results (and reports) from all of these trials would be included in the NDA. There was a minor problem. A large number of the older, completed trials had NOT been reported. It was a very arduous task to locate the data, sift through the information, and then compile the data into a coherent report. The team working with me did an absolutely outstanding job performing this task. However, it shouldn't have been required at that stage. While it's a natural tendency to want to finish the study, get the answer, and then to move on to the next study, it is the sponsor's legal and ethical obligation to report the study in a timely fashion. The words of advice: *Don't let the report completion process slip.*

4.5.3 Monitoring and Safety

4.5.3.1 Monitoring

The purpose of monitoring, of course, is to protect subjects, to make sure the data are accurate and verifiable, and to be compliant with good clinical practices.

Monitors must be appropriately trained, knowledgeable, and experienced. They must be familiar with all the information related to this particular product (such as information contained in the Investigator's Brochure), this particular protocol, all the ramifications of this protocol, and be fully trained and knowledgeable about the whole gamut of GCPs as well.

Monitoring has two sets of meanings. There is someone called a clinical monitor. Typically, that's a physician. The physician might be within the sponsor's organization or might be a contract person from a contract research organization. The person will have significant medical experience and can make medical judgments during the course of the study (or development program).

The sponsor will also have monitors with titles such as Clinical Research Associate or Clinical Research Assistant (CRAs) who have training and experience interacting with sites, reviewing site documents, reviewing case report forms, verifying data, etc. On rare occasions, in very small companies these two functions might be performed by the same individual, but almost always they are different. The terms can be confusing nonetheless.

The CRA tends to be the person who is the primary communication pathway between the sponsor and the investigator. The monitor ensures that the investigator and their staff will also have appropriate qualifications, resources, training and the appropriate documentation will be in place — as discussed previously. Likewise, CRAs ensure that the investigator and the site staff are performing appropriate trial functions adequately. They will also make sure the appropriate subjects are enrolled into the study. They will ensure that records and reports are adequate. Are they legible? Are they signed? Are they initialed in the appropriate location? The CRA will also make sure that the CRFs are accurate and that they correspond to the support or primary documentation. This monitor is also going to ensure that the adverse events are appropriately reported.

Monitoring at trial sites also follows a set of procedures that are in accordance with a set of standard operating procedures. Those

SOPs are either from the sponsor or from the CRO. Typically, if the sponsor is a relatively small company and a CRO will be hired, an SOP will need to be in place that tells how that process is going to be handled. The SOP will describe the identification and contracting process (in general), how the sponsor will monitor them, and assure that they are doing good quality work. Obviously, if monitoring is being handled internally, the SOP needs to be appropriate for the internal monitoring activity.

The monitor (CRA) needs to prepare a report after every visit. Certain visits may occur telephonically or via teleconference; in either case, they need to be documented. Visits which occur in person at the site will report on all activities conducted during the visits and, importantly, any issues identified or actions required as a result of the visit. Whether the sponsor or the contractor conducts the visit, it is important that issues are resolved and pending actions are closed out and documented. Monitoring activities are discussed in Chapter 9.

4.5.3.2 Medical Expertise

The medical expertise of the monitors (CRAs) needs to be appropriate — for the protocol and the program. That description is not terribly specific, but it means the monitors have enough expertise so they could recognize appropriate (or inappropriate) information in a CRF. For example, could they determine that this was an adverse event? That's an element of their training and their qualifications. Monitors tend to have a nursing background, but could be pharmacists or practical nurses. In some countries around the world, it is common that CRAs are physicians. At the least, the CRA needs to be someone who has an adequate medical background to understand all the information contained in the CRF.

4.5.3.3 Adverse Drug Reaction Reporting

In the case of adverse events, the CRA needs to understand and follow all the procedures necessary to report the particular reaction, as appropriate, to investigators, institutions, IRBs, or regulatory authorities.

Reports prepared by them will also comply with applicable regulatory guidelines; in particular, ICH guidances. Typically, the monitor will not be reporting those adverse events directly to the institution, IRB, regulatory authorities, etc. Nevertheless, the person who does the monitoring will understand what the procedures are within the company or within the CRO. Safety aspects of clinical trials are covered in greater detail in Chapter 11.

4.5.4 Investigational Products

The sponsor is also responsible for all aspects of investigational products. This includes providing appropriate information to investigators and vendors, ensuring that manufacturing, packaging, and labeling are handled appropriately and that the supply and handling of investigational products is carried out in a responsible manner.

Information typically contained in the Investigator's Brochure should contain sufficient safety and efficacy data from the non-clinical and clinical studies to support the trial. As we have noted previously, the IB is the source (in summary format) of all the information known about the drug up to the present time.

The manufacturing, packaging and labeling must conform to good clinical practice and good manufacturing practice. The sponsor determines and implements all packaging and storage requirements. For example, if the study will include sites in Alaska and Florida and it's going to occur in the dead of winter or the heat of summer, the sponsor needs to understand how the product will behave in those particular circumstances. Study supplies also need to be packaged, labeled (as an investigational product) and coded (for randomization) appropriately.

Remember that supplies are not provided to the investigational site until ALL approvals and documentation are in place at that site. The sponsor provides all handling and storage instructions AND ensures they are actually followed. In turn, the CRA will review the site's handling of drug supplies, as well as the maintenance of records regarding receipt and distribution of study supplies.

4.5.5 Quality

In quality assurance and quality control, QA / QC, the sponsor will ensure that the adequate and appropriate SOPs are in place.

If the sponsor is a very small company, much of the quality work will be contracted. If that is the case, the appropriate SOPs dealing with the contract process must be in place. In turn, the contracted organization, the CRO, will have SOPs in place dealing with their own activities.

A QC review at each stage ensures reliability of data. Remember that part of the principles of good clinical practices is that the sponsor is not only protecting the patient, but also ensuring that the data are useable and reliable.

The purpose of auditing is really to evaluate the trial conduct and the compliance with the protocol, with the SOPs in place, with good clinical practices and with regulatory requirements. The auditors need to have training in this particular aspect of evaluation. They need to be independent of the clinical trial itself; they need to be independent of the data collection system. They also need adequate training and experience. All of that needs to be documented. The procedures that the auditors use will be included in an additional set of SOPs. The auditors are also guided by the importance of a particular trial to the submission. They will make sure that whatever findings they have are documented and that whatever issues related to noncompliance have been identified will be rectified.

It is the sponsor's responsibility that appropriate access to records is available to the appropriate individuals. The sponsor ensures that the investigator and the institution provide direct access to the source data for monitoring, audits, IRBs, ethical committees and regulatory inspection. The sponsor will verify that individual subject consent is in place prior to the participation of each and every subject. Even if the sponsor hires a CRO to do all of this work, the sponsor needs to make sure that these records are in place.

4.5.6 Finances

The sponsor also has very specific responsibilities regarding the financial aspects of the trial.

Compensation to subjects and investigators is an important aspect of any trial. The topic includes insurance and indemnification. These items need to be identified in the informed consent. The informed consent will discuss how adverse events are going to be handled from a financial standpoint, whatever the costs of treatment for trial-related injuries might be and how that kind of compensation will occur. Details related to each of these items need to be in place at the beginning of the trial. This is the responsibility of the sponsor. There are a couple of options: The sponsor might compensate all expenses above what is covered by insurance or perhaps cover all expenses associated with the trial (generally the latter is better). In any case, the sponsor commits to following and caring for the patients until resolution of the adverse event.

All financial aspects between the sponsor and the investigator and the institution are documented in written agreements. Today, this is particularly important because of new concerns (in all regions of the world) related to disclosure of all real and perceived conflicts of interest. Items such as ownership of the sponsor company's stock by the investigator, or that he or she is receiving an outrageous amount of money for "consulting", are the types of things that are deemed "conflicts of interest". Whatever the arrangement turns out to be, it needs to be documented.

If there is premature termination of the trial, the sponsor needs to make sure that it has provided notice and explanation to the appropriate agencies involved in the trial, including other investigators at other institutions, regulatory authorities and, then, the appropriate IRBs and ethical committees.

Chapter 5

Essential Documents

5.1 Definition and Description

What are essential documents? The identification and specification of these documents is contained in the ICH Guidance E6. **Essential documents** are those documents that enable evaluation of the conduct of a trial, enable evaluation of the quality of the data produced during the trial, assist in successful management of the trial and serve as source information during audits by either the sponsor or regulatory authorities.

A logical way to organize these documents is by the timing of their use in relationship to the actual trial conduct. Cleverly, that timing might be: before, during or after the trial.

Table 5-1 identifies the documents that need to be in place at the investigative site *prior* to enrollment of any patients including a core set that includes the IB, the protocol, the consent form, agreements between sponsor, site, institution, CROs, etc., curriculum vitae (CVs) of all individuals involved, laboratory normal values and laboratory certifications (if local laboratories are used).

In addition, approvals may need to be collected from central regulatory authorities; this is more commonly required in locations such as Eastern Europe or India.

Table 5-1. Essential Documents: Before Trial Start

Essential Documents
Core Documents
Investigator's Brochure
Protocol
Informed Consent Form (ICF)
Agreements (institution, sponsor, CROs, etc.)
Curriculum Vitae
Laboratory Normals
Approvals
Institutional Review Board / Ethical Committee
Regulatory authorities (as required)
Product
Procedures for handling of product
Procedures for decoding blinded product
Initiation
Site Initiation Visit Reports

Source: ICH Guidance, E6

Additional documents in this group include handling instructions for the product and randomization procedures, as well as a copy of the study initiation report from the CRA.

The next set of documents includes those pertinent to the *on-going* trial. (See Table 5-2.) If any of the core documents have been updated during the preparation time for the trial, the updates need to be included. In addition, approvals by the IRB for any of those amended documents must be filed. Records of shipping labels and certificates of analysis for the product are also included. Other on-going activity documentation included in this group is copies of periodic monitoring reports, signed consent forms, appropriate source documents and screening and enrollment logs.

Documents required *following* completion of the study include copies of the final report to the IRB or IEC and the clinical study report. The logs documenting accountability or destruction of investigational product and the close-out monitoring report are also maintained. (See Table 5-3.)

Table 5-2. Essential Documents: During Trial Conduct

Core Documents (Updates, revisions, or amendments)
Investigator's Brochure
Protocol
Informed Consent Form (ICF)
Agreements (institution, sponsor, CROs, etc.)
Curriculum Vitae
Laboratory Normals
Approvals for Any Amended Documents
Institutional Review Board / Ethical Committee
Regulatory authorities (as required)
Product
Shipping Labels
Certificates of Analysis
Ongoing Activities
Monitoring Reports
Signed (and witnessed) ICFs
Source Documents
Enrollment Logs

Source: ICH Guidance, E6

Table 5-3. Essential Documents: After Trial Completion

Final Documents
Final Report to IRB / EC
Clinical Study Report
Informed Consent Form (ICF)
Product
Accountability
Destruction / Return
Close-Out Activities
Monitoring Report

Source: ICH Guidance, E6

5.2 Investigator's Brochure

Details regarding the Investigator's Brochure (IB) are also identified in ICH Guidance E6. (See Table 5-4.)

The first section of the IB is an introduction, which includes a compilation of the relevant clinical and non-clinical data up to this point in time. The IB always has a date and version number so the most recent version can be identified. The IB is designed to provide investigators, and anyone else who's involved in the trial, information to understand the rationale for the study or the key features of the protocol or a perspective of the overall development plan. The IB also contains a section the current risk:benefit assessment — in the context of data currently available.

The title page includes the sponsor's name, identity of the product, the date and edition of the IB, confidentiality statement, and signature page. The first portions of the IB include the Table of Contents and the summary highlighting information from the sections below.

The introduction includes the chemical name and pharmacologic class. Importantly, it also identifies the rationale for the study and for the indications for the overall development program. There may be sub-sections addressing conduct of specific studies, but, in general, the summary addresses the rationale for studying the product in the specified indications, e.g., hypertension, and what types of patients are most likely to benefit from the use of this product in the course of the development program.

It is important to put the current study(ies) in the context of the overall development plan. However, keep in mind there is the fine point at the edge of information here. A certain amount of information is essential for those investigators, site staff, IRBs and regulatory agencies considering the study. On the other hand, recognize that, on occasion, those IBs do show up in other people's hands. As a result, you don't necessarily want to include an excruciating amount of detail in here about all of the plans for future study.

Physical, chemical and pharmaceutical properties are next. Particularly important in this section are the storage instructions. This is especially true if the product has any particularly unique qualifications or requirements in terms of storage.

Then, non-clinical studies are reported. The section includes all relevant information. A lot of this shows up in tables, instead of being

Table 5-4. Investigator's Brochure: Structure

Introduction	
Title Page	
1.	Table of Contents
2.	Summary
3.	Introduction (Product, Rationale, General Study Plan)
4.	Physical, Chemical and Pharmaceutical Properties and Formulation
5.	Non-Clinical Studies
6.	Effects in Humans
7.	Summary of Data and Guidance for Investigator

Source: ICH Guidance, E6

descriptive — it's easier to reference and locate. The implications of the studies reported in the tables needs to be discussed. In particular, whatever extrapolations might be made to use in humans should be specified and discussed. Sections on the non-clinical pharmacology, pharmacokinetics and product metabolism, at least as understood to this point in time, as well as the toxicology are included.

The known effects in humans are then discussed. If this is the first clinical trial with the product, this information may not be known specifically. If information is known about similar pharmacologic classes or other drugs out of the same pharmacologic class, it should be discussed here. The discussion should also include information relating to similarities and differences between the subject product and other known reference products. Such data may or may not be available; if they are available, they enable the investigator to put the product in perspective as he or she pursues the investigation.

Once data in humans are available, a list of the information available needs to be included. This is the section where all adverse events (AEs) known to date are delineated. Whether adverse events fit into the category of "expected" is determined by whether or not the event is listing in the IB. If it is listed, it is "expected"; if it isn't listed, it is "unexpected" at this stage. Presence or absence of an AE in the IB thus determines some reporting requirements for events occurring in on-going studies.

The key thing to understand and put in perspective is that all information related to safety, including prior human experience AND non-clinical information, is relevant to current and future human trials. ALL available information needs to be included. Some may be summarized and some may be presented in more detail, but in all cases the information must be put in perspective.

It is recognized, of course, that product development is an evolutionary process. New information will become available and evaluable at later points in time. As PK and product metabolism, safety and efficacy data become available in humans, the data need to be included. Occasionally, marketing experiences in other countries for this particular product might be available. It is, perhaps, a bit more challenging to keep that up to date, but it is of importance to current investigators.

Next, there is an overall summary of the data and the guidance for the investigator. The data are summarized in a form that enables investigators participating in the trial to have a clear understanding of possible risks and adverse reactions, as well as benefits. The overdose section addresses the sequelae if a patient takes too much of this particular drug. Whatever the extent of that experience might be needs to be summarized.

Again, if it's the first experience in man, the only reference point may be non-clinical data; then, summaries and extrapolation are based on that experience. It is possible, even by the time of submission of the NDA, that no one has taken an extremely large dose of the drug. In that case, guidance for usage (and treatment of overdoses, if they were to occur) is based on information available.

Drug interactions also need to be investigated; particularly for drugs that might be used frequently with this medication. Based on the experience available, these data need to be included in the section as well. At the end, the risk:benefit advice is based on analyses of information available at the time the IB is written. The current study(ies) is(are) placed within the context of the available information.

5.3 Clinical Trial Protocol

The numbers in Table 5-5 correspond to the sections of a Clinical Trial Protocol, as defined in document E6. We will cover the information relatively quickly, since the document is easily available and can be used as a reference for all clinical studies.

The cover sheet contains the protocol title, name and address of the sponsor, whoever is monitoring, whatever medical expert you have utilized to put together this protocol, the investigators' names and institutions, the physician who is responsible for monitoring this particular protocol, and then what other organizations or institutions (e.g., laboratories) that are performing services related to the study.

The background information is a section that details the name of the investigational product and includes a summary of the relevant non-clinical findings. The details of this will be in the IB, but there is also a summary here, within the Clinical Trial Protocol.

The protocol also includes a discussion of the benefits and risks, based on information available to date. (It, as we have noted above, might include only non-clinical information or might include a large body of clinical information.) Whatever data are available need to be summarized in this portion of the protocol.

The population to be studied is specifically identified. For example, the overall indication for the drug might be hypertension, but this protocol might focus on a specific subset of that population.

A description and a justification for the route, dosage, dosage regimen and treatment periods for the particular drug or therapeutic modality to be employed in the study need to be described. In addition, a statement needs to be included stating that conduct of the study will be in compliance with "The Principles of the Declaration of Helsinki" and with "good clinical practices".

The trial objectives and the purpose of the study need to be clearly stated so that those conducting it fully understand the sponsor's objectives.

The next section has to do with trial design. Endpoints need to be succinctly and clearly defined. I will offer up my bias, which is that there should only be one primary end point in a study. I have seen too many study designs with multiple "primary" endpoints. Secondary endpoints may be multiple in number. Exploratory endpoints, I don't believe should — as a general rule — be identified as primary or secondary endpoints. The exceptions to this statement are those situations with small populations and a "new" endpoint is being evaluated. (See discussion in Section 6.3.1.)

One other item to note is the current tendency to structure studies with a primary endpoint that is a composite endpoint, meaning that several observations are combined into a single endpoint. There are sound statistical approaches to such studies, but the approach requires component endpoints that are relatively straightforwardly measurable.

Description of the trial design with a diagram in the protocol, is, I find, very helpful for people to understand the flow of a particular study. The protocol specifies those measures taken to avoid bias, for example, blinding and randomization, as well as the description of the treatment, dosage, and schedule.

The protocol will also contain whatever stopping rules will be in place for individual subjects or for the trial itself.

The inclusion / exclusion criteria section must clearly specify the patient population or subject population that is to be included in this particular study. It is important to clearly define the people to be included and to clearly define the people to be excluded.

Table 5-5. Clinical Trial Protocol: Structure

1.	General Information
2.	Background Information
3.	Trial Objectives and Purpose
4.	Trial Design
5.	Inclusion / Exclusion of Subjects
6.	Treatment of Subjects
7.	Assessment of Efficacy
8.	Assessment of Safety
9.	Statistics
10.	Access to Source Data ./ Documents
11.	Quality Control / Quality Assurance
12.	Ethics
13.	Data Handling and Record Keeping
14.	Financing and Insurance
15.	Publication Policy
16.	Supplements

Source: ICH Guidance, E6

Withdrawal criteria also need to be clearly specified. For example, if conducting a study of hypertensive patients with concomitant moderate renal failure, the protocol might specify the change in serum creatinine that would require withdrawal of the patients from the study.

The next section of the protocol — treatment of subjects — describes in detail how patients will be treated and on what schedule. Screening visits are defined as well as on-study and follow-up visits. Descriptions include procedures conducted, including laboratory studies and other assessments, medication changes dictated by the protocol, etc. All the details related to each of those visits need to be specified.

Concomitant medications that are permitted and excluded need to be clearly identified. For example, if the study focused on pain relief, use of acetaminophen or aspirin would most likely be restricted or excluded. That needs to be pretty clearly specified in this section. Procedures for monitoring compliance will also be specified and spelled out here.

Within the "Assessment of Efficacy" section are the clearly defined efficacy parameters and the process for their analysis in the study. For example, a decrease in blood pressure of 10 mmHg or more at the second or fourth or eighth or tenth or 15th visit could be defined as "efficacy". The protocol defines what the measurements are, when they will be taken (and how they would be taken), and the analysis process for determination of efficacy.

The protocol also defines the analytical approaches for safety. Thus, the protocol will specify the safety parameters to be incorporated into the study, as well as the methods and timing of those assessments. Duration of follow-up of adverse events needs to be specified and whatever procedures for reporting. These are some of the things that we talked about previously in terms of adverse event profiling.

The statistics and the statistical methods in the protocol need to be described, including any interim analysis, how that interim analysis is going to be conducted, how it's going to be triggered and so forth. It is also useful to include the number of subjects expected / permitted at each investigative site. Generally, when specified, a range of minimum to maximum is specified.

The level of significance of the key statistical assessments is also specified. The statistical section could also describe a trial termination process based on either an interim analysis or review by a Data Monitoring Committee.

Procedures for dealing with missing data need to be specified ahead of time, before you ever enroll a patient in the study. The whole set of procedures involved in the statistical analysis plan, or SAP, needs to be specified before a single patient has enrolled in the study and before you begin to look at any of the data.

The final list of items contained in the Clinical Trial Protocol specifies certain procedures and activities. These include: access to source documents, procedures for QC / QA, ethics, data handling, record keeping, financing, insurance, and publication policy.

The final section includes supplements required to support the information and procedures described in the protocol. These sections tend to be boiler plates. So, once you have done these for your first protocol, they tend to not be modified very much. As you build a new protocol (or the second, third, fifth or the tenth protocol in your series), you tend to just use these as a boiler plate format and import them into the new protocol (reviewing them periodically, of course, to make sure they are up to date).

Chapter 6

Introduction to Clinical Trial Design

Figure 6-1 describes, in schematic fashion, the process of clinical trial planning. It starts with a strategy meeting which, taking into account the science around the disease and the therapeutic target and current medical practice, leads to a definition of study objectives, study variables and a study design. Next, an analysis plan is developed that is then incorporated into the clinical trial protocol, translated to draft case report forms and then ultimately incorporated into a database. Finally, the trial is actually conducted, and data management and statistics contribute to analysis of the data and a final study report.

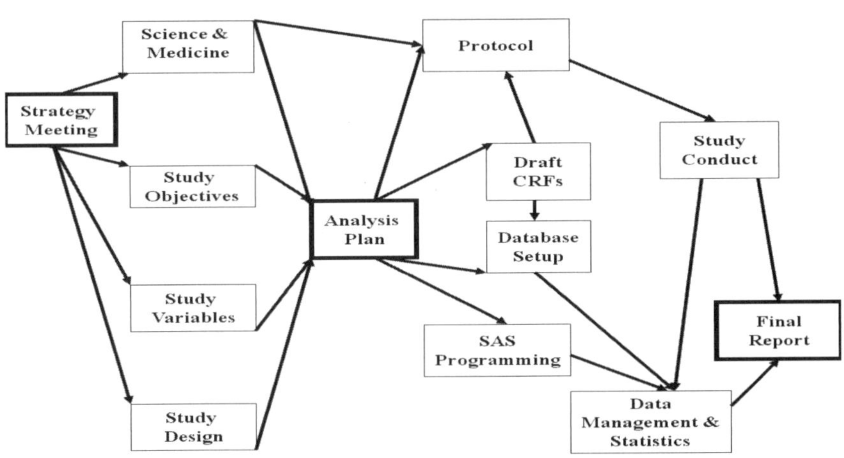

Figure 6-1. Process of Clinical Trial Planning

What is a **clinical trial**? My definition (there are many out there) is: *A clinical trial is a systematic investigation and evaluation of the effect of a device, diagnostic, or pharmaceutical product (chemical or biological) in a population of normal subjects or patients.*

I admit it is a pretty broad definition, but, to my mind, it encompasses everything from first-in-man studies all the way to studies of marketed products. I believe the principles and considerations we will discuss in this chapter (and book) apply to that full range of studies.

The document that is created to serve as a guide and recipe manual for the conduct of the trial is called a **protocol**. (See Section 5.3.) Table 6-1 contains a detailed outline of a clinical trial protocol as specified in ICH E8, "General Considerations for Clinical Trials".

A good protocol (there is no such thing as a "perfect" protocol) enables the conduct of an effective clinical trial — one that provides an interpretable answer at its conclusion.

Table 6-1. Clinical Trial Protocol Outline

I. General Information
- A. Protocol Title and Protocol Identifying Number
- B. Sponsor information
- C. Monitor information
- D. Investigator information
- E. Laboratory information
- F. CRO information

II. Background Information
- A. Description of the Investigational Product
- B. Summary of Non-clinical Studies Relevant to the Trial
- C. Assessment of Potential Risks and Benefits
- D. Description of , and Justification for, the Route of Administration, Dosage, Dosage Regiment and Treatment Period(s)
- E. Compliance Statement
- F. Description of the Population to be Studied
- G. Relevant Literature References

Table 6-1. Clinical Trial Protocol Outline (continued)

III. Trial Objectives and Purpose
- A. Efficacy
  - 1. Primary question and response variable
  - 2. Secondary questions and response variables
  - 3. Subgroup hypotheses
- B. Safety

IV Trial Design
- A. Description of the primary and secondary endpoints
- B. Trial design (including schematic diagram of design, procedures, and stages
- C. Description of measures taken to minimize / avoid bias
  - 1. Randomization
  - 2. Blinding
- D. Trial treatment
  - 1. Dosage and dosage regimen
  - 2. Description of dosage form, packaging, and labeling
- E. Duration of subject participation (sequence and duration of trial periods)
- F. Stopping rules / discontinuation criteria
  - 1. Individual subjects
  - 2. Parts of trial
  - 3. Entire trial
- G. Accountability procedures for all investigational products
- H. Procedures for maintaining and breaking treatment randomization codes
- I. Description and identification of source data

V. Selection and Withdrawal of Subjects
- A. Subject inclusion criteria
- B. Subject exclusion criteria
- C. Subject withdrawal criteria and procedures for withdrawal

VI. Treatment of Subjects
- A. The treatment to be administered
- B. Medication(s) / treatment(s) permitted (including rescue medication) and not permitted before and / or during the trial
- C. Procedures for monitoring subject compliance

VII. Assessment of Efficacy
- A. Specification of the efficacy parameters
- B. Methods and timing for assessing, recording, and analyzing or efficacy parameters

Table 6-1. Clinical Trial Protocol Outline (continued)

VIII. Assessment of Safety
- A. Specification of safety parameters
- B. The methods and timing for assessing, recording, and analyzing safety parameters
- C. Procedures for eliciting reports of —and for recording and reporting — adverse events and intercurrent illnesses
- D. The type and duration of the follow-up of subjects after adverse events

IX. Statistics
- A. A description of the statistical methods to be employed, including timing of any planned interim analysis (analyses)
- B. The number of subjects planned to be enrolled
  1. In multicenter trials, the numbers of enrolled subjects projected for each trial site should be specified.
  2. Reason for choice of sample size, including reflections on (or calculations of) the power of the trial and clinical justification
- C. The level of significance to be used
- D. Criteria for the termination of the trial
- E. Procedure for accounting for missing, unused, and spurious data
- F. Description and justification of procedures for reporting any deviation(s) from the original statistical plan
- G. Selection of subjects to be included in the analyses (e.g., all randomized subjects, all dosed subjects, all eligible subjects, all evaluable subjects)

X. Direct Access to Source Data / Documents
- A. Statements ensuring agreement that the investigator(s) / institution(s) will permit trial-related monitoring, audits, IRB / IEC review, and regulatory inspection(s), providing direct access to source data / documents

XI. Quality Control and Quality Assurance

XII. Ethics
- A. Description of ethical considerations relating to the trial

XIII. Data Handling and Record Keeping

XIV. Financing and Insurance
- A. (If not addressed in a separate agreement)

XV. Publication Policy
- A. (If not addressed in a separate agreement)

XVI. Supplements
- A. (As needed)

Source: ICH E8 "General Considerations for Clinical Trials"

6.1 What is the point?

What are you trying to accomplish with the trial? Again, the key advice is, "Begin with the End in Mind".

ALL TRIALS SHOULD START WITH A QUESTION. In some cases the question might be relatively simple; in other cases the central question might involve several dependencies and be relatively complex. However, in all cases there needs to be a question! In even the simplest observational trial, there must be a legitimate question to answer.

In the terms of a formal statistical approach, the question is ultimately posed as an **hypothesis**. It starts like this . . . At the beginning of the trial-creating process, the first utterance might actually be an "I wonder" statement — "I wonder if treatment A is better than treatment B in patients with hypertension." As the observational possibilities of the statement are pondered, however, the statement needs to be rephrased as a question — "Is treatment A better than treatment B in patients with hypertension?" Then, in considering the magnitude of the challenge, the next consideration will be the manner in which the question is approached. How can the question be tested? What things need to be considered to obtain an accurate answer? Well, actually, what is an accurate answer? It all gets pretty complicated pretty quickly.

To help us frame the process, that is, in order to design an effective clinical trial, one needs to ask (and then answer) several questions:

- What are the objectives of the study?
- What are the endpoints of the study?
- What criteria should be used to select subjects or patients for the study?
- How many subjects or patients should be included so that the study will have enough power to detect a clinically relevant result?
- How do we protect the subjects or patients entering the study?
- How will we monitor the safety of all subjects involved in the study?
- How sure are we of the results (are there factors that might obscure the interpretation of the data)?

6.2 Types of Clinical Trials

There are many different types of clinical trials, employing different approaches to generate the best data possible.

Indeed, there are pros and cons to every trial design. Parallel group studies (see below), for example, are the most common design, but because of inter-individual variability, tend to require larger numbers of patients in order to achieve statistical significance. Alternatively, cross-over designs (see below) enable comparisons within individuals, i.e., the individual serves as their own control, and generally require fewer subjects, but are operationally a bit more difficult and tend to be utilized in smaller studies (e.g., Phase 1 pharmacokinetic studies) or studies in very limited populations (e.g., orphan diseases).

One element of categorization is determining the trial's context within the development process (see Chapter 1). For example, whether the trial is part of Phase 1, Phase 2, Phase 3, or Phase 4 will affect its objective and design.

Another categorization of trials is whether or not a control is used, i.e., the trial may be **controlled** or **non-controlled**. In controlled trials, patients are allocated to receive one treatment or a reference (control) treatment. (See 6.3.3.2, below.) The process by which this occurs is called **randomization.** But the randomization process might take several forms: **Parallel group** (randomized evenly or in some other fixed allocation sequence; subjects will receive only one of the choices of treatment, analytical comparisons are between treatment groups), **cross-over trials** (subjects receive, in a randomized order, each of the treatments, thus serving as their own control), or **factorial trials** (multiple combinations are possible for a single patient).

The number of clinical trial sites (study centers) employed in the study to recruit patients also enables trial categorization. Trials may be conducted at a **single center** (investigative site) or in a few or many centers. Studies conducted at single centers tend to be early (e.g., Phase 1) and may involve specialized studies, e.g., pharmacokinetics. Studies with dozens or even hundreds of centers (**multicenter** studies) tend to be later in the development program and are frequently those involved in registration (i.e., pivotal studies, Phase 3) or in post-marketing (Phase 4).

The goal of the study, i.e., what it is designed to prove, also permits discrimination. A **superiority study** aims to show that a new drug, device or treatment is more effective (or safe) than the comparative treatment (e.g., placebo or a current standard of therapy). An **equivalence study** is designed to prove that two drugs or treatment approaches have the same clinical benefit or safety (or that the difference between the two is clinically insignificant). A **noninferiority trial** aims to show that the effect of a new treatment cannot be said to be less significant than that of the current treatment, but — within the limits of the trial design — neither can it be said the two treatments are equivalent.

The goal or intent of the study may also be looked at another way. The first study designed to evaluate the new treatment in the relevant patient population might be labeled as **exploratory**. It is designed to, for the first time, estimate the effectiveness and safety of the new treatment compared to some standard of therapy. On the other hand, later trials designed to duplicate or reinforce earlier studies are described as being **confirmatory**. These are generally studies conducted in Phase 2b and Phase 3 portions of development programs.

6.3 Key Aspects of Trial Design

It is now decision time. In order to design and implement the best trial possible (still not "perfect"), key choices must be made. While not an exhaustive list of things to consider, this group of topics will deal with and highlight many of the key issues:

- Endpoints
- Patient Selection
- Duration of Treatment
- Sources and Control of Bias
- Randomization
- Blinding
- Sample Size and Power

6.3.1 Clinical Trial Endpoints

What is an endpoint?

Above, we described the process of developing a study question. We also suggested it needed to be reframed as an hypothesis. The statement of the hypothesis is usually in a form called a **null hypothesis**. The null hypothesis states that there is no statistically significant difference between our two treatments of interest with respect to the endpoint chosen.

A clinical trial endpoint is a measure that, following analysis, enables a decision regarding acceptance or rejection of the null hypothesis. Endpoints can be designated as primary or secondary. Primary endpoints measure outcomes that will answer the most important question being asked by the clinical study (as framed around that endpoint). The size of the trial (discussed below) is determined by the power needed to detect a difference in this endpoint between study groups (or study arms). Secondary endpoints frame additional questions around other factors (measures) that are important in understanding the primary results.

While the concept of a primary endpoint assumes a single measurement and question, in fact multiple measurements may contribute to the endpoint itself. These endpoints are referred to as **composite endpoints**.

Examples of endpoints includes the following types of observations: change in a biomarker (e.g., a straightforward laboratory test, such as a blood glucose or cholesterol level, or a more specialized test, such as troponin, a marker of heart failure) or a measurement (e.g., blood pressure); change in symptom pattern (e.g., less cough, fewer headaches, less joint pain), change in quality of life (e.g., better self image, more self sufficient, improved exercise capacity and an ability to work), changes in outcomes (e.g., death, death from a specific cause), and improvements in utilization of resources (e.g., less cost, fewer supplies, fewer hospitalizations).

6.3.1.1 Choosing Endpoints

Since the whole point of the study is understanding the effects of treatments or interventions on endpoints, it is important to select them well. It is important to ensure that:

- The endpoint selected is clinically meaningful and has an understood relationship to the disease,
- Changes in the endpoint will provide an answer to the main question of the trail,
- Measurement or assessment of the endpoint is feasible and practical and can be determined in the same way in all subjects, and
- The endpoint occurs often enough (or can be measured often enough or with enough precision) to enable creation of a sufficiently robust statistical analytical process.

6.3.1.2 Surrogate Endpoints

A surrogate endpoint in a clinical trial is some kind of laboratory measurement or physical sign used as a substitute for a clinically-meaningful endpoint that, in turn, directly corresponds to a patient's well-being, ability to function or survival. Therefore, changes induced by a therapeutic intervention on a surrogate endpoint are expected to reflect changes in a clinically-meaningful outcome measure. A couple of examples will help reinforce the point. Blood pressure, cholesterol levels and hemoglobin A_{1C} have well established links to mortality. Therefore, regulatory agencies will accept beneficial changes in those markers as "surrogates" for measuring mortality. In addition, today, many anti-cancer therapies are approved based on surrogate markers. On the other hand, BCR-ABL is a protein marker associated with chronic myelogenous leukemia (CML). There are data suggesting that reductions in the quantitative measurements of the marker are associated with improved outcomes. However, as of this writing, quantitative changes in the marker are not an approvable endpoint for new therapeutics for this disease.

Use of a surrogate endpoint (or multiple endpoints) can significantly reduce the sample size needed for a trial. In turn, the duration and cost of performing a clinical study are decreased. Surrogate endpoints are particularly useful when conducting Phase 2 screening trails to identify whether a new intervention has an effect, since the surrogate endpoint might be affected long before the intended therapeutic endpoint. If the surrogate is directly linked to the "real" outcome measure, an early, relatively inexpensive, "yes or no" might be obtained from the trial.

It is crucial to remember that surrogate endpoints are only useful if they are a good, direct predictor of a specific clinical outcome. If the relationship is not direct or is not consistently observable or determined, it really has limited utility.

6.3.1.3 Composite Endpoints

As noted above, composite endpoints result from combining other individual endpoints into a single measure. Composite endpoints can be useful because, in general, a smaller number of patients might need to be included (reflecting an increase in the number of observed outcome events) and a more comprehensive evaluation of a treatment may be permitted across more than a single category of outcome. However, the utility of composite endpoints may be limited because of the assumption that each of the components is directly and equally related to the outcome as well as being equally beneficial. That assumption might prove to be incorrect.

There are some limitations to composite endpoints. In a composite endpoint of multiple outcomes, we make the assumption that avoiding any one outcome is of as equal importance as avoiding any other outcome. That is, however, rarely the case. The second assumption made when using composite endpoints is that all individual outcome measures are related to the disease process and are equally meaningful. This also may not always be true.

6.3.2 Patient Selection

Next, the subject population is identified. If the trial is an early (Phase 1) study, the population is likely to be healthy volunteers. If the study involves patients, the specific population of interest needs to be clearly identified. For example, is it broad, e.g., all hypertensive patients, or narrow, e.g., class 4 congestive heart failure patients? The types of patients studied throughout the development program (an in each protocol, in particular) will ultimately drive the product labeling, so the inclusion / exclusion criteria become extremely important. Not uncommonly, programs will start with a relatively tightly defined population, and the population will broaden as development is continued.

For an individual clinical trial, a sample of patients is taken from the population of interest. For example, our program is interested in the efficacy of a potential antihypertensive compound. To determine its potential efficacy and safety, it is important that the trial or trials be conducted with patients reflective, in some direct way, of the larger population of patients who would be expected to use (and benefit from) the new compound. Thus, it is important that the selected sample truly reflect the population it represents, and that the eligibility criteria are not so restrictive that they hamper recruitment or that the results would not be generalizable to the larger population.

6.3.2.1 Source of Patients

One of the first considerations when designing a clinical trial is to establish the source of patients to be recruited into the trial. The sample size and the clinical outcomes to be measured will need to be balanced against practical considerations. Phase 1 and early Phase 2 trials are generally small and may involve a special population (e.g., normal subjects for early pharmacokinetic studies; patients with severe forms of cancer or congestive heart failure) and may be conducted in specialized centers (e.g., Phase 1 units or Clinical Research Centers). Larger Phase 2 and all Phase 3 studies, however, generally require large numbers of patients and the source of clinical sites is usually quite varied, i.e., academic centers, group practices, and solo practitioner settings.

In order for recruitment to be successful, it is important that patient selection is performed at centers with enthusiastic investigators (also with diligent and motivated study coordinators). While you want investigators who are enthusiastic, you also want investigators who are knowledgeable in the appropriate clinical area as well as knowledgeable in clinical trial conduct. In addition, the investigators also need to have access to the facilities and equipment that are required to perform the trial-related tests and measurements. (Remember the Investigator Responsibilities discussed under GCPs in Chapter 4.) It is also important to understand whether the center and / or the investigator have competing studies, particularly in the same indication. It is partly a question of internal resources and it is partly a question of availability of appropriate patients.

As one works with potential investigators, it is important to query in some detail whether the investigator / center has the appropriate patients and has enough of them. One technique to employ to better understand the potential pool of patients is to request that investigators complete a potential patient log. In these days of HIPAA (see Chapter 7), one cannot request actual patient names; however, one can request the investigator to list potential patients on a log sheet that identifies inclusion / exclusion criteria and requests that the investigator either check boxes or record values for specific patients who would then qualify for the study. In general, sites can do this from their computerized records. Recruitment is discussed in more detail in Chapter 9.

6.3.2.2 Eligibility Criteria

In order to reduce bias and variability and increase the power of the clinical trial, it is essential that well-developed eligibility criteria be in place. As noted above, though, the criteria should not be so restrictive as to limit enrollment.

A key aspect of eligibility criteria is to define patients with the disease or condition to be evaluated. It is essential that standardized criteria be utilized to describe the population under consideration. Occasionally, one might try to identify previously-undefined subsets of patients who might benefit most from the therapeutic intervention. However, the core population should fit a definable group. Remember, the goal is to write the product label. If the population isn't definable, the product will not be approved for marketing.

To make this a bit more concrete, let's look at an example. Imagine that one is looking at a new therapy for congestive heart failure. There are standardized definitions of the syndrome of congestive heart failure and there are descriptions that enable one to define severity. As the clinical trial is defined, then, one starts with the standard definitions. They might be refined for the study to identify a specific subset of patients, e g., left-handed patients who also have chronic renal failure. While perhaps a bit unusual, the target population would be quite clearly described. Eligibility criteria are usually based on:

- Patient characteristics (e.g., left-handedness),
- Diagnostic test results (e.g., serum creatinine greater than 3 mg/dL),
- Disease duration (in this case it might be the duration of the respective diagnoses of congestive heart failure and chronic renal failure), and
- Disease severity (e.g., class 4 congestive heart failure).

As the trial is described, the patient who would enter the study must meet ALL the inclusion criteria and have NONE of the exclusion criteria. In general, the simpler such criteria, the more useful and easily applied they are.

The goal is to recruit and enroll as many of the patients with the disease as possible, but to exclude those with co-morbidities that might affect outcome and those in whom safety might be compromised.

If particular tests or measurements are to be performed as part of the screening process to determine eligibility, it is essential that the results of these investigations are available prior to the time at which the patient would be randomized into the study.

6.3.2.3 Disease Status

It is important to select patients who are ill enough to improve with the intervention and to exclude those patients who are too ill to participate or who have specific characteristics that would render them unlikely to respond to the intervention. These aspects of selection should be clearly defined in the protocol.

6.3.3 Duration of Treatment

Duration of treatment and observation are the next elements to consider. As a general statement, patient exposure is related to the extent of toxicology data available. So, if toxicology data are available for 28 days, the longest human exposure can be 28 days. If exposure data are available through 90 days, the longest corresponding human exposure is 90 days. When chronic toxicology studies are completed (6 and 9 months, respectively), then human exposure can extend as long as product stability allows and regulatory authorities permit.

In general, pharmacokinetic and pharmacodynamic studies are of short duration, e.g., up to approximately 2 weeks. Phase 2 studies permit multiple doses, generally out to about 12 weeks. Finally, Phase 3 studies generally entail longer dosing intervals (except, for example, antibiotics for short-term administration) and are in line with ultimate product labeling and intended usage.

Duration of treatment also must be explored from the standpoint of time to expected effect. For example, if an antihypertensive drug would be expected to have its effect within two weeks, the trial duration — looking at efficacy — must be at least two weeks in duration. But then, you need to understand whether that is initial effect or maximal effect. Furthermore, one needs to understand whether the effect is sustained or goes away over time (i.e., tolerance develops).

Furthermore, one also needs to understand timing to possible side effects — the safety side of the equation. For a treatment with expected long-term usage, e.g., antihypertensive, the chronic use side effect profile is very important to understand. So, the studies need to be long enough to reasonably understand that profile.

From the sponsor's perspective, though, time is also related to money. While the duration of the studies needs to be "long enough" to provide adequate risk:benefit information, longer studies have increased trial costs and extended trial durations, and are accompanied by higher operational costs of the company itself. Particularly for smaller companies without on-going revenue sources, this can be an exceedingly important consideration.

6.3.4 Sources and Control of Bias

Bias is an "opinion or feeling that favors one side in an argument or one item in a group or series, predisposition, prejudice". In clinical research, bias is defined as systematic distortion of the estimated intervention effect away from the truth, caused by inadequacies in the design, conduct, or analysis of a trial. The actions or decisions that result in bias don't have to be intentional to, nevertheless, have a profound impact on the conduct and interpretation of the trial.

There are potentially many sources of bias in clinical trials. We will focus on only six: lack of controls, selection bias, bias in study management, observer bias, bias in post-study analysis exclusions, and publication bias.

6.3.4.1 Lack of Controls

I would suggest that all studies have controls built into them — even early Phase 1 and proof of concept studies. On the efficacy side, you just can't tell about benefit if there are no controls. And on the safety side, study results — i.e., adverse events — need to be put into perspective and that is difficult without any controls.

Controls can either be placebo or active. Placebo controls are those without active treatment, although the subjects still receive a pill, capsule or solution that is similar to what is received by the active treatment group. Active controls are generally the current standard of treatment, but might also be other active approved therapies. Use of active controls tends to be later in development when there is evidence of success already in hand and the study is designed to do comparisons for positioning. However, some trials will incorporate both active and placebo arms as a way of verifying the efficacy and safety profile of the investigational drug. Use of active controls is common in the European Union, even in early phase studies.

Placebo Effect

We have known for years about the "placebo effect" or the "placebo response". Much of the discussion was started by Henry Beecher's paper in the Journal of the American Medical Association in 1955 entitled "The Powerful Placebo". It turns out, over the years, that the placebo response in clinical trials appears to be getting more powerful, i.e., placebos seem to be working better! Another factor, particularly in the era of multi-national trials, is that the placebo response has a relationship to cultural factors — patients who still have faith in or respect for physicians (simply because they are physicians), e.g., in Eastern Europe, seem to have a more powerful placebo response than patients in other environments, e.g., the United States. It is also known that the color of tablets (active or placebo) may be associated with a response. For example, yellow pills seem to do better as anti-depressants, green pills relieve anxiety better, and white works well as an antacid. Hopefully, a not-well-publicized project at the NIH (the Placebo Response Drug Trials Survey) will contribute valuable information to our understanding.

Although I generally don't like the idea of non-controlled studies — and open label studies fit into that category — I will acknowledge there are occasions when an open label exploratory study is the correct approach. An example might be in a study addressing a rare disease that has easily identifiable endpoints. A small exploratory study might be conducted to see if there is enough of a clinical response to justify proceeding further.

Historical controls are used rarely except, again, in situations where patients are very hard to come by. In that case, the historical control may be the only option. Regulatory agencies have approved orphan drugs in which historical controls have served as the reference population. (However, before embarking on such a program, detailed discussions with the regulators are absolutely essential.)

6.3.4.2 Selection Bias

Selection bias occurs if a systematic difference exists in the way in which study subjects are enrolled (either accepted or rejected) into a trial or in the way that treatments are assigned to those enrolled. In turn, of course, such actions can influence the trial results and their interpretation.

One tool employed to reduce selection bias is **randomization** — the random allocation of individuals to one or more study groups. (See discussion on randomization below in Section 6.3.5.) An additional mechanism employed to reduce that selection bias is **stratification** into subgroups. This involves an additional step or steps in the randomization process to control variables that ultimately result in an even allocation of participants into the respective groups.

Selection bias can also be introduced if a highly-selected group is enrolled into a trial (e.g., in order to ease the demands of patient recruitment or to minimize inter-subject variability). The treatment effect in such a group might well be different from that in the actual (target) population of interest.

6.3.4.3 Bias in Study Management

The interpretation of a randomized controlled trial relies on the assumption that any differences in outcome are the result of either chance (whose effects can be quantified) or of inherent differences between treatments. This assumption, however, is invalid if the treatment groups are not handled equally with regard to all of the study procedures. This means that all protocols should clearly define all study-related procedures so that all subjects are handled the same way.

6.3.4.4 Observer Bias

When knowledge of the treatment assignment (by participants already recruited into a trial, investigators, or persons who analyze and report trial results) leads to systematic distortion of the trial conclusions, this is referred to as **observer** or **ascertainment bias**.

The best way to avoid observer bias is to conduct trials as blinded studies. (See Section 6.3.6, below.) This means that some or all of those involved in the study (study participants, investigators, assessors, etc.) are unaware of the treatment assignment. We addressed selection bias above. Selection bias is minimized by protecting (shielding) the randomization code before the subjects are randomized until the treatments are administered; observer bias is avoided by protecting the randomization code until all treatments have been administered to all patients.

6.3.4.5 Bias Created by Post-Randomization Exclusions

It is common that during the trial some data are missing — patients may miss visits, laboratory samples might be lost or mishandled, etc. Decisions regarding these data and how to handle them are an integral part of the trial and the data analysis process. Inappropriate handling of the data can also lead to bias. Two types of analyses are typically done after a clinical trial (other types of analyses are also done, but these are almost always done): a per-protocol analysis and an intent-to-treat analysis. A **per-protocol analysis** excludes all subjects who were non-compliant, have missing data, or were unable to complete their assigned treatment. In an **intent-to-treat analysis**, all the study participants enrolled in the trial are included in the analysis, regardless of whether they completed the trial according to the protocol or not.

There are (at least) two variations on the intent-to-treat analysis. One, I will label **"pure" intent-to-treat**. This analysis would include any patient who has been randomized — whether or not they received medication. The other is generally called **safety analysis** (although it is used for efficacy as well). This analysis includes all of those patients who were randomized AND received at least one dose of study drug. And, finally, there is at least one variation on the safety analysis, also cleverly called **safety analysis**, that includes patients who were randomized AND received at least one dose of study drug AND have at least one measurable data point after receiving the drug.

Whatever groupings are going to be used for analysis really MUST be identified at the beginning of the trial, preferably in the statistical analysis plan, to avoid issues of bias.

6.3.4.6 Publication Bias

Historically, positive (results suggested efficacy for the treatment under study) trials would be written up and submitted for publication. Negative trials, however, were seldom published, leading to a bias in the literature. The Food and Drug Administration Modernization Act of 1997 (FDAMA) required sponsors to submit limited data to the www.clinicaltrials.gov database. Then, the Food and Drug Administration Amendments Act of 2007 (FDAAA) required all clinical trials of drugs, biologics and devices — except Phase 1 clinical trials — to be registered at www.clinicaltrials.gov AND required that the results of trials that would serve as the basis for an efficacy claim and post-marketing trials must also be posted. While neither of these pieces of legislation guarantees that negative results will show up in medical journals, at least the results will be in the public domain.

6.3.5 Randomization

The **randomized controlled trial** (RCT) is considered the "gold standard" for testing the efficacy of medical treatments. A fundamental assumption that forms the basis of the RCT is that patients in different groups are similar for characteristics that may affect the outcome of the trial. In many cases, the selection of characteristics is straightforward — age, sex, severity of disease, types of prior treatments, etc. However, there are other times when the impact of certain characteristics is not known or, at least, well known enough at the time of initial trial design. It is the intent of the RCT to minimize the impact of all such characteristics on the outcome of the trial as much as possible. The importance of this activity is apparent at the time of review of the data. At that time, the data are analyzed under the assumption that effects seen in the trial are related to the treatments applied (or random error). Any process or activity that negates this assumption invalidates the trials (and calls into question the ethics of conduct of the trial).

Randomization is a process of assigning patients to treatment groups by a code which renders the treatment group unknown to the patient, the investigator or to others actively involved in the clinical trial. The randomization sequence can be used to vary the number of patients in each of the treatment arms and can also be arranged in such a way that stratification of patient groups is accomplished to achieve better group balance.

To achieve the intended balance, patients are randomly assigned to each treatment modality in such a way that the characteristics of interest are divided equally among the treatment arms of the study. The randomization sequence, or code, is typically generated by a statistician using a computer-generated random-number table.

A further requirement of randomization is that it must not be predictable by the person assigning patients to the treatment modalities, otherwise there is a chance the groups will contain bias. To prevent this, certain methods of "blinding" are used so that patients and staff are not aware of the treatment distribution. Blinding is described in a bit more detail below.

There are several different randomization methods. Among them are: simple randomization, block randomization, stratified randomization, and minimization or adaptive randomization.

Simple randomization is a procedure that makes each new treatment allocation without regard to those already made. The method is easy to implement and is unpredictable (in terms of bias for study or sponsor personnel). In general, the top level of randomization is very close, e.g., treatment A vs. treatment B. However, there might be small inequities in that randomization that could be more important in smaller studies. More importantly, though, it is blind to potentially important subgroup differences that might be known at the time of study start.

A **block randomization** procedure also uses randomly-generated assignments. However, the block constrains the process so that the assigned groups are periodically balanced. For example, the randomization code in a study with a block size of six would ensure that with every six subjects entered into the study, the respective groups would be balanced. The size of each block of allocations must be an integer multiple of the number of treatment groups. For example, if there are three treatment arms, block sizes could be 3, 6, 9, etc.

The balance forced by blocking is especially important in long-term trials if recruitment is slow, the type of patients recruited in the trial might change during the enrollment period, and there is a consideration of stopping the study early for safety or efficacy reasons.

Stratified randomization elaborates on the balance defined in the block randomization process. The primary randomization allocates patients into the respective treatment groups, e.g., Treatment A and Treatment B. A secondary randomization is then employed to balance a set of predetermined important factors that might impact the prognosis (response) of the patient in the trial. Examples might include age, gender, diagnosis of diabetes, severity of illness, etc.

Adaptive randomization attempts to minimize the differences between treatment groups based on selected prognostic factors. As a general rule, the trial starts by randomizing patients in a standard fashion. Then, after a predetermined number of patients, the number of patients who have been allocated to the respective treatment groups are examined to determine the presence or absence of a specified characteristic in the respective groups. If the distribution of the characteristic of interest is unequal between the treatment groups, the next several patients would be specifically assigned to the group with the smaller numbers.

6.3.6 Blinding

At each stage of a clinical study, there are a variety of individuals who can introduce bias. These individuals include patients, principal investigators, physicians, surgeons, sponsors, and other healthcare professionals, local study coordinators, core labs reporting on study samples, trial statisticians, and committees such as the Data Monitoring Committee (DMC). A blind for the study may be imposed on any of these groups.

Blinding is placed in effect so that subjects only (single blinding), subjects and investigators and investigator staff (double blinding), or subjects, investigators, and sponsors (triple blinding) are unaware of the treatment group to which a subject has been assigned.

There are four primary categories of blinding:

- Unblinded studies (open),
- Single blinded,
- Double blinded, and
- Triple blinded.

An **open or unblinded study** is, as described, conducted with all participants aware of the various treatment arms and comparisons. Occasionally, unblinded trials make sense in exploratory evaluations to determine whether a larger, more expensive, more complicated trial might be worth the effort and expense. Such trials tend to be more "realistic" in the sense of reflecting true clinical practice. However, there are numerous disadvantages. Adverse events in such trials, somewhat surprisingly, tend to be underreported. There is also a bias toward efficacy unless the measurement is a lab test. But even with lab tests, patients might alter behavior in a way that might secondarily affect the test.

Single-blinded studies involve a blinding of patients, but all other personnel involved in the trial (e.g., investigators, sponsors) are aware of the therapeutic interventions. Such a trial is relatively simple to conduct, but is again plagued with several potential problems. The investigators or other study personnel might consciously or unconsciously influence the patient's perception of efficacy or safety, i.e., over-reporting efficacy and under-reporting of adverse events. The investigators might also treat the patients somewhat differently, likewise affecting results.

In **double-blinded studies** neither the patient nor the investigator is aware of the assigned treatment intervention. The double-blinding process only avoids biases if the investigator and the patient can remain blinded throughout the study. Thus, all other aspects of the trial — how patients are treated and handled, the medications, the assessments — are the same between the two groups. If the sponsor is unblinded, however, certain biases may again creep into the study. The sponsor might, consciously or unconsciously, affect the investigator's assessment of efficacy or his or her handling of an adverse event.

Triple-blinding a study means that the patient, the investigator and all members of the sponsor's team are unaware of the interventions in the respective treatment arms. Some of the advantages to this scheme are obvious, such as the reduction in potential biases in conduct of the study. Triple-blinding, though, adds significantly to the complexity of the study and requires specific procedures to be put in place to ensure adequate safety monitoring of the study. Triple-blinding tends to be standard procedure in large trials that are to be included in regulatory submissions as the primary evidence for marketing approval.

6.3.7 Sample Size and Power

The sample size of a clinical trial is the number of subjects that are to be enrolled in the study. Choosing the right sample size is critical for a trial and is based on certain key assumptions. The first assumption regards the size of the benefit in the measurement of interest we anticipate with the experimental treatment compared to the standard (or placebo) treatment (also called the "expected treatment effect"). The second major assumption is the amount of certainty we wish to have with which to capture the treatment benefit (also called the "power of the study").

As a general statement, the larger the sample size, the better the power with which to detect a treatment effect. Therefore, smaller differences between treatments can be detected. (The converse is also true: The smaller the trial, the larger the difference between treatments is required for detection.) The calculation employed to determine the sample size is also dependent on the trial design.

There are several factors that determine sample size.

Expected event rate in the control arm

The expected summary measure of the primary endpoint (such as event rate) in the control or standard treatment arm of the trial is determined through a review of available information. In general, this is obtained by a review of relevant literature regarding the disease or the specific aspect of the disease being studied. Rarely, because of the unusual nature of the aspect being studied or the rarity of the disease itself, the literature might be insufficient, and interviews of experienced investigators might be required. In that case, it is best to validate the endpoint with the regulatory agency.

Smallest effect worth detecting

Again, a review of the literature is the place to start this evaluation. However, in determining the true clinical significance of the purported benefit, consultations with experts are quite useful. There is little utility in conducting a study to produce a result that is judged by experts in a field as being clinically meaningless.

Significance level for rejecting the null hypothesis

The significance level is the level at which we will reject the null hypothesis that there is no difference in the treatment effects. We are trying to determine in essence the "false positive" threshold for assessing the study. When evaluating only a segment of the entire population who have the particular traits of interest, there is a possibility we can generate data that support rejecting the null hypothesis (i.e., the data suggest the treatment arm do indeed have a difference in the marker of interest) when, in fact, that difference does not exist in the entire population. We need to understand there is a possibility of a false positive result and set a threshold to minimize the likelihood of its occurrence. This is called **Type I error** and is usually set at 0.05 or 5% and is designated by the Greek letter α. To decrease that likelihood further and to lower that threshold, more patients are required in the study.

Power of the trial

The reverse of the situation relates to the possibility that we fail to detect a difference in our study when there actually is (or would be) a difference in the population. In this situation, we have obtained a "false negative" result. We should be rejecting the null hypothesis ("there is no difference"), but we, in fact, fail to reject it based on the data obtained. We would like to reduce the chances of this error, called **Type II error**, also. In general, this threshold is set at 20% and is represented by the Greek letter β. The power of the study (that is, the ability to find a significant difference if it exists) is 100%-β or 80%. If one wishes to be more certain, the threshold for Type II error can be reduced to 10%. Therefore, the power is increased to 90%. Increasing the power of the study is also generally accompanied by an increase in the number of study subjects.

Study design

This is actually a complicated topic requiring an experienced statistician. However, as a general rule, a parallel study with the standard treatment group (or placebo) acting as the control will require more subjects than a study employing the subjects as their own control (e.g., assessments done before and after an intervention). In the parallel study, different ratios of randomization (e.g., 1:1, 2:1, 5:2, etc.) will also impact the sample size (non-equal randomizations require more subjects). The assessment variable (e.g., whether categorical or continuous) will also affect the sample size.

Subject drop-out rate

The statistical approaches to these calculations result in numbers of subjects who will have data available for comparison, i.e., those subjects who complete the study. Essentially all studies will have subjects who drop out for various reasons (lack of interest, travel, side effects, perceived lack of efficacy, etc.). Those subjects need to be added to the number expected to complete to determine the number of subjects who should be enrolled in the study. In addition, the screen failure rate also must be considered in these calculations. One does not want to get to the end of a trial and find oneself a few patients shy of a hoped-for statistical result.

6.4 Statistics

Statistics will be considered in more detail in Chapter 10 in the discussion of data flow. Here we will briefly consider statistics within the context of designing an individual clinical trial.

Understand that high quality data, convincing data and clear-cut analyses that address both efficacy and safety are needed, and that the statistical choices made affect the trial design strategy. Thus, ask the right questions of all involved in the trial design — particularly the clinician and the statistician.

It is at this point that **probability** raises its ugly head. Although the term might not have been considered since college (or even high school), its importance is in understanding the likelihood that the population selected for any trial (or the whole program) represents the population actually targeted with your therapeutic agent.

Next, we need to consider the measurement scales to be employed in the study. Are the questions being asked answerable with a simple yes or no? Or is the scale graded, such as a pain scale? Or are the outcomes related to survival? Perhaps the assessments of interest are improvements or worsening of a disease. For simple questions, e.g., "Yes vs. No", measurements (e.g., blood pressure), or laboratory results (e.g., blood glucose), the value recorded on the case report form is pretty straightforward (though the analyses might still be complex). However, what about assessments of quality of life, or disease progression / remission, pain assessments? In those cases, some kind of scale or composite assessment can be used. Then, there are concerns about validation of that scale or assessment. Is it broadly applicable for the disease or for many diseases? Does the language in which it is administered matter? Do the instructions matter? What are the appropriate assessments of "change" or "improvement", i.e., how will you analyze the data? The approvability of the drug may ultimately rest on the ability to provide detailed, concrete and convincing answers to these questions.

At the start of a program, some things are known and some things aren't. Occasionally, a trial needs to be conducted to understand the right things to measure. That solution, though more time-consuming and more costly, is generally better than the "coin-flip" approach and enables generation of a robust null hypothesis.

A useful document in approaching the design of clinical studies is the ICH Guidance E9: Statistical Principles. (Again, refer to Chapter 10 for a more complete discussion.)

6.5 Adequate and Well-Controlled Studies

In the United States, regulations state that approval for marketing will be based on "substantial evidence" of effectiveness and that evidence will be collected in "**adequate and well-controlled trials**" [21 CFR 314.126]. There are several key characteristics of "adequate and well-controlled studies" to consider in the design of a single trial or an entire program.

- The protocol contains a clear statement of the study objectives.
- There is a summary of the proposed analytical methods within the protocol and there will be a statement in the final report of the actual analytical methods used.
- The study design permits a valid comparison with the control used in the study to provide a quantitative (read "statistical") assessment of the drug's effect. The control within the trial could be placebo or an active drug or both.
- The endpoints used — the methods of assessment of the subjects' responses (for both efficacy and safety) are clear, well-defined and reliable.
- The study report provides sufficient details of the study design, conduct, and analysis to understand what was actually done and to have confidence in the report of the data.
- The patients were assigned to treatment and control groups in a way that minimized bias (generally handled through the randomization process).

- Adequate measures were employed to minimize the potential bias of the subjects, observers and recorders of data, and those involved in the data analysis (generally covered under the subject of blinding).

6.6 The Trial within The Program

Re-look at Figure 6-1. The process described in the figure, on a larger scale, also represents the process for creating a development plan for the product as a whole, as discussed in Chapter 3.

Chapter 7

Protection of Human Subjects

7.1 Background

As we consider development of new therapies for human diseases, testing in animal models is much less than predictive of either efficacy or safety in humans than we would hope. Thus, if the societal goal is improvement of the population's health, evaluation of newer and better medications and treatments will need to be performed in human subjects.

Human Subjects and Clinical Research

There is no question that there is a great need for research using human subjects. Society demands improvements in health care and will demand newer and better medications and treatments. Desperate patients may even seek unproven remedies if research results are not forthcoming.

http://www.irbforum.org/

The need to study new therapies in humans brings to light a whole series of issues, not the least of which is the protection of the rights of each individual who might possibly participate in clinical research.

The purpose of clinical research, as noted in Chapter 6, is to answer scientific questions that enable us to understand disease processes and treatment outcomes. By answering two deceptively simple questions ("Is it safe?" "Is it effective?"), we can deliver new therapeutic modalities to the marketplace. Unfortunately, the answers to these questions are frequently less than perfectly clear. Then, as we summarize all of the information available to us, we are left with trying to assemble a benefit:risk ratio that, in turn, enables us to decide whether or not to put that therapeutic modality into the marketplace. Deciding on the balance of benefit to risk really is a balancing act and always requires consideration of imperfect information. We study hundreds or perhaps a few thousand patients whose job it is to stand in as surrogates for the millions of patients who might potentially be exposed to the therapy we are developing.

Thus, to get better therapies, we need to conduct studies in people.

However, the conduct of trial in human subjects has, unfortunately, produced some very un-pretty moments along the way.

Development of research ethics standards has, in almost all cases, resulted from some nasty triggering event. (See Table 7-1.)

Table 7-1. Research Ethic Milestones

Trigger Events	Ethics Milestones
Syphilis Study	Nuremberg Code, 1947
Nazi Experiments	Amendments to FD&C Act, 1962
Radiation Experiments	Declaration of Helsinki, 1964
Thalidomide Tragedy	Belmont Report, 1979
Milgram Study	HHS / FDA Regulations, 1981
Beecher Article (1966)	CIOMS Guidelines, 1982
	Common Rule, 1991
	National Bioethics Advisory Committee

7.1.1 Nuremberg Doctors Trial

In 1946, in Nuremberg, Germany, physicians and other individuals who had participated in activities during World War II on behalf of the Nazi regime were subjects of a trial. (A powerful movie, *Judgment at Nuremberg,* captures the essence of these trials.) The Nazi regime exploited human beings by forcing them to participate in research without their consent. The mortality rate of these wartime research studies was at least in the range of 25% to 30%, though likely much higher.

The Nuremberg trials were heard from December 1946 to August 1947. There were 23 defendants. 20 of those were physicians. They were charged with murder, torture and other atrocities committed, ostensibly in the name of medical science. Fifteen of the 23 were found guilty and seven were sentenced to death. The judgment that came out of the Nuremberg trials included a set of standards known as the Nuremberg Code, an ethical yardstick, which has been subsequently used as a foundation for much of what's happened in the intervening 60-plus years regarding protection of human subjects.

A 1997 article highlighted the enduring importance and significance of the Nuremberg Code. The article sets the Nuremberg Code within the historic context of its chronologic occurrence and within the context of the cascade of events that transpired subsequently. In addition, it also specifically identifies the key elements in the Nuremberg Code and highlights the ramifications of the results of the Nuremberg trials. (Shuster, 1997)

7.1.2 Thalidomide Tragedy

Thalidomide was a drug approved as a sedative in Europe in the late 1950s. Because the drug was believed to have no side effects — at least initially — it was particularly used for women who were pregnant.

The FDA did not approve the drug — at least at that time. However, samples were sent to doctors in the United States.

Doctors distributed these samples without telling the patients the drug was experimental. Physicians frequently use samples for patients

— usually the goal is to assess the utility and safety of a drug in a patient before writing a full prescription (thus saving the patient money and some risk). All that logic goes away, of course, when the drug isn't approved for marketing in the US. In this case, the drug was not approved in the US, but was, nevertheless, dispensed by the physicians to the patients as a sample.

By 1961, data were available both in the United States and around the world that thalidomide was harmful to the fetus by interfering with normal development of arms and legs, and produced a condition called phocomelia (i.e., very short arms and legs). (Subsequent studies determined the risk to be in a very narrow window, about two weeks, in early gestation. It is currently on the market for treating leprosy.)

There are several problematic aspects to the thalidomide story: importation of the drug into the US despite the fact that it had not been approved; failure to identify to patients that this was an experimental drug; and failure to understand or appreciate the significance of the data that had been accumulating about these side effects.

The cumulative impact of these issues led to passage of the Kefauver-Harris Amendments to the Food, Drug and Cosmetic Act in 1962. These amendments represented the first US statute that required subjects to be informed of a drug's experimental nature and to consent to receiving that drug before starting participation in the research study.

7.1.3 Milgram Study

Stanley Milgram was a psychologist at Yale University who conducted a study focusing on the conflict between obedience to authority and personal conscience. (Milgram,1963) He examined the justification for acts of genocide offered by those accused in the Nuremberg trials; their defense was often based on obedience, that is, that they were simply following the orders of their superiors.

In the experiment, Milgram recruited a group of people, so-called "teachers", who were actually the unknowing subjects of the experiment. They were asked to administer an electric shock of increasing intensity to a learner, "a student", for each mistake the learner made during the course of the experiment. The fictitious story given to the teachers — the actual subjects — was that the experiment was exploring the effects of punishment for incorrect responses on learning behavior. The teacher was not aware that the learner in the study was actually an actor merely indicating discomfort as the teacher increased the electric shock and the "shocks" themselves weren't real.

When the "teacher" asked whether electric shocks should be given, they were encouraged to continue. Sixty per cent of the "teachers" obeyed orders to punish the "learner" to the very end of the 450-volt scale. No "teacher" stopped before reaching 300 volts. In fact, some of the subjects appeared to lose consciousness.

There were times when the worried "teachers" questioned the experimenter, asking about responsibility for any harmful effects resulting from shocking the "learner" at such a high level. When assured that the experimenter was assuming full responsibility, all of the "teachers," — *all of them* — seemed to accept the response and continued shocking, even though some were, obviously, extremely uncomfortable in doing this.

The study raised many questions about how subjects could bring themselves to administer such heavy shocks. Over and above the questions about basic human nature, there are many questions about the conduct of basic clinical research. Indeed, the role of deception of human subject research continues to be debated even today. Current Federal regulations continue to allow some deception in research, but only in limited conditions and only with IRB approval.

I would suggest you review a book, *People of the Lie*, written by Scott Peck and published in the early 1990s. Peck's focus in the book really has to do with the study of evil in general and how it occurs in society. One particular example relates to the collection of events that occurred in Vietnam in what was called the My Lai Massacre, in which a group of innocent villagers were killed by American soldiers. The logic offered as a defense was: "I was simply following orders." While certainly interesting in the context of human evil (refer to Nuremberg, above), that discussion also has a fundamental relationship to the structure and conduct of clinical trial research.

7.1.4 Beecher Article (1966)

Henry Beecher was an anesthesiologist who died in 1976. In 1966, he published an article in which he reviewed 22 various research studies that had been published. (Beecher, 1966) In the course of his review, he cited serious ethical problems in this group of 22 studies. There were really more studies to consider; he only selected these 22 because he judged them to be the worst of the dozens of studies he reviewed.

It took a fair amount of time, but about 16 years later, the Council for the International Organization of Medical Sciences, CIOMS, published a set of internationally-applicable ethics guidelines for biomedical research involving human subjects.

7.1.5 Tuskegee Study (Syphilis Study)

A study, called the "Tuskegee Study of Untreated Syphilis in the Negro Male," started in 1932 in Macon County, Alabama. A group of 600 subjects were originally recruited into the study; 201 served as controls and 399 were infected with syphilis. In 1947, penicillin was as the treatment for syphilis, yet it was not offered to the subjects. Although originally planned to last 6 months, it was not until 1972 that the study was discontinued and treatment was extended to subjects; two years later, treatment and care was extended to the subjects' families. (In 1997, President Bill Clinton apologized on behalf of the nation.)

Ultimately, in 1974, the National Research Act required regulatory protection for human subjects. The legislation also created a commission to evaluate the protection of human subjects and make recommendations. This resulted in a very important document, called the **Belmont Report**, which was published in 1979. Thus, nearly 35 years from the end of World War II, including the aftermath leading to the promulgation of the Nuremberg Code, a report finally addressed detailed ethical principles as they apply to the treatment of research subjects.

Early in the 1980s, there were changes in the Health and Human Service and Food and Drug regulations related to the protection of human subjects. In 1991, what's called the **Common Rule** (http://www.hhs.gov/ohrp/policy/#common) was passed. The Common Rule was then applied to all federally-funded programs involving human research and the protection of subjects involved in such research.

7.1.6 Human Radiation Experiments

Unfortunately, that is still not the end.

In 1993, the Albuquerque Tribune reported on a series of experiments detailing the administration of plutonium to unknowing subjects (including pregnant women) at several major academic medical centers, including UCSF and UC Berkeley, ostensibly to study the effects of the atomic bomb. (A book was subsequently published by Eileen Welsome, entitled The Plutonium Files: America's Secret Medical Experiments in the Cold War.)

In one "study," 73 disabled children in a Massachusetts school were spoon-fed radioactive isotopes along with their morning oatmeal. In an upstate New York hospital, Manhattan Project doctors injected an 18-year-old woman with plutonium; she believed she was being treated for a pituitary disorder. At a Tennessee prenatal clinic, 829 pregnant women were served vitamin cocktails, which, in truth, contained radioactive iron as part of their prenatal treatment.

As a result of these studies, an advisory committee on human radiation experiments was formed in January of 1994. This, ultimately, led to the National Bioethics Advisory Commission in 1995.

In its Charter, the National Bioethics Advisory Commission (www.bioethics.gov) states, "It is essential that the research community come to value the ethics of research as central to the scientific process." So, our research needs to take into account the protection of human subjects and to value the ethics of such research in the human population.

7.1.7 Key Lessons

As we review these events over the last 50 to 60 years, there are several points for us to remember. First, investigators bear the ultimate ethical responsibility for their work with human subjects. Society entrusts to those investigators the privilege of using other humans to advance scientific knowledge. When any investigator, and I will emphasize ANY investigator, ignores the basic principles of ethics, the entire research community suffers.

Compliance with ethical principles, I very strongly believe, should be seen as the right thing to do because it helps protect the rights and the welfare of subjects of human research, not simply because somebody says you have to do it, or that it is a box to check off before you can get your study approved.

7.2 Declaration of Helsinki

As noted in Section 4.2, The World Medical Association published a very important document — The Declaration of Helsinki — in 1964. (The most recent version (Seoul) is dated 2008 and is available on the World Medical Association website, http://www.wma.net/en/30publications/10policies/b3/index.html.) The importance of The Declaration of Helsinki is that it serves as the foundational document for the ethical conduct of human research.

The document makes a distinction between research with therapeutic intent and research for scientific knowledge. This is, of course, appropriate for our consideration as we structure clinical research programs. The document also specifically calls for oversight of research.

7.3 The Belmont Report

The Belmont Report (http://ohsr.od.nih.gov/guidelines/belmont.html) was issued in 1979. It is such an important document, with implications for all that we do in clinical research, that we will spend some time discussing its key points.

It's a report of the National Commission for the Protection of Human Subjects of Biomedical and Behavioral Research. It was very directly triggered by the Tuskegee study and, obviously, by some of the other events that we recounted previously.

It called, specifically, for oversight of research by institutional review boards (the first time such an entity was specifically created to directly review clinical research).

In the original Declaration of Helsinki, the mechanism for that oversight of clinical research was not really well defined. It was stated as a principle, but the mechanics weren't created at that point. The Belmont Report went on to define and describe what would be, and what has subsequently become, an institutional review board. Outside the US, these are typically called institutional ethical committees, but the concept is fundamentally the same.

Key principles in the Belmont Report include a respect for persons, a concept called beneficence, and a concept called justice.

7.3.1 Respect for Persons

The initial principle is that individuals have **autonomy**. They can control their own body, their own health, their own well-being and, as a result, they can control their participation in clinical studies — providing their own consent.

In addition, there are people with diminished autonomy ("**vulnerable populations**") who are entitled to increased protection. Several groups fall into this category, such as prisoners and children, subordinate individuals, and individuals who are decisionally-impaired.

Prisoners

The potential advantages for the prisoner who might consider participation in a clinical trial should not interfere with his or her ability to make a voluntary choice, fully weighing the risks and the benefits of the study. For example, parole decisions should be unaffected by either their participation or non-participation in a clinical research study.

In addition, the Belmont Report specifically states that the risks of participation for a prisoner would be just as acceptable to non-prisoner volunteers, as well and that randomization processes need to be "fair" in their allocation to treatment or non-treatment groups. Furthermore, all such subjects would receive adequate medical care while participating in the study.

Children

It would seem logical, of course, that if the potential study subject is an infant, or a six- or seven-year-old child, he or she is not really in the position to fully understand the potential — good or bad — of their participation in a clinical research study. Someone, a surrogate, is needed to take care of the patient and protect their autonomy; that person typically is the parent (although it could be a court system).

Children cannot legally grant consent, but they must provide active assent. There is a point in age when a subject needs to be able to say, "Yes, I am willing to participate" and then, in addition, their parents or their guardians concur and provide the legal consent.

The assent process, however, obviously must be appropriate for the age, maturity or psychological state of the child. A child of 16 may be quite mature, may be very psychologically stable and able to provide their own assent. On the other hand, another child of the same age may not be in the position to fully understand and fully comprehend the implications of trial participation.

45 CFR 46 specifically recommends that all 7- to 12-year-olds provide separate assent. Accordingly, the explanation to the seven-year-old would need to be in language appropriate for that individual and would need to include — as for others participating in trials — what he or she would experience as a result of participating in this particular trial, what the potential benefits might be, and what the potential risks might be. In the US, it is also important to understand regulations in the individual states, since they do vary.

Subordinate individuals

Subordinate individuals are individuals, such as college students, employees, and persons in position of lesser power or status.

For example, imagine you are a student, in a psychology program, or other kind of research program, and someone in authority (e.g., your professor) suggests (strongly states) that you need to participate in this study if you want to get a good grade in this class. The Belmont Report directly addressed that situation as being a significant ethical issue. The incentives to participate must not be "unduly influencing".

Essentially, the same thing is true for employees within companies. We know that, in many biotechnology companies, there is an opportunity for employees to participate in clinical research. However, the process for participation must be non-coercive and the risks must not outweigh the perceived benefits.

Each of these individuals — no matter whether they are students, employees, or other persons in positions of lesser power or status — must be able to decline, without prejudice, participation in a particular clinical trial.

Furthermore, confidentiality must be maintained for all self-disclosures that occur as a result of participation in a research project. Thus, college students and employees must be reassured that completion of questionnaires with information required for the conduct of the study does not prejudice their situation; that is, the data will be used for no other purpose than the research in question.

Decisionally-impaired

Impaired mental status is considered under the label "decisionally-impaired". This population would also, at least, include those with psychiatric illness, substance abuse, and various metabolic disorders that might influence the ability to have logical thought processes, hence impairing an individual's ability to make a decision.

7.3.2 Beneficence

The Belmont Report defines **beneficence** as "unconditional goodwill and compassion for all those within reach of our influence; quality of being kind; charitable, doing good". Beneficence as a concept really implies "to do no harm". In addition, it implies a maximization of the benefits to the participants in a clinical study and, likewise, a minimization of the risks. Thus, consideration of the benefit:risk ratio falls under the concept of beneficence.

Harm (risk) may be considered first as likelihood (statistical probability of occurrence of some harm) and second as a degree of severity (the harm might be mild, moderate, or severe in degree).

Benefit is anything of positive value related to health or welfare that might occur as a result of the person's participation in a particular study. That benefit could be, in fact, some improvement of their health or the psychological benefit from knowing that someone (the study personnel) is paying attention to them, or even the psychological benefit of believing their participation in the study will benefit others. All of those things might be considered in the term "benefit".

7.3.3 Justice

Under the **justice** aspect, the issues identified in the Belmont Report include subject selection and a fair distribution of the risk and the benefit, as well as a fair application of the inclusion and exclusion criteria for participation in the study.

"Individual justice", as applied in the Belmont Report, requires fairness in application of the inclusion and exclusion criteria.

"Social justice", on the other hand, requires drawing a distinction between classes of people who are asked to participate in the study, yet treating them equally. Within the research community, we have not done this with consistency. For example, recent work has looked at multiple clinical trials conducted over a number of years and found that the approach to women, blacks and children has been skewed, resulting in a consistent under-representation of such individuals in certain trials. For many years, the "pivotal" studies for most drugs were performed in large populations of white men. Yet, when the drugs were marketed, they were prescribed for much broader populations. As a result, the profile of benefits:risks was not adequately evaluated in the pertinent patient populations.

Subsequently, there have been some specific attempts to rectify these situations. Now, for example, clinical research conducted under NIH guidelines requires certification that the proposed study will adequately include representative samples of men, women, various ages, and various racial populations (unless there is a really good reason to exclude some group, e.g., women in a prostate cancer study). Those requirements are logical extensions of the findings and conclusions in the original Belmont report.

The distribution of the risk and the benefit must be fair among all participants or potential participants. This means, simply, that you have not stacked the deck in terms of which people are potentially exposed or not exposed to certain risks or benefits in the course of the study.

Injustice, the flip side of this particular term, would mean that entitled benefits — those things that would normally accrue to anyone who would participate in this particular study — are denied to individuals, groups, or subgroups of individuals without a good scientific, ethical, statistical reason. Or, to turn it around in terms of adverse events, burdens are imposed unduly on certain segments of the study population.

The Belmont Report addressed each of these items and, as a result, served as the foundation for what could be called "modern" approaches to protection of human subjects.

Despite the fact that this report was published in 1979 and that a number of regulations have been promulgated since then to put these principles into effect, there are, as can be observed from the radiation experiment studies, still problems.

I believe firmly that it is crucially important for all of us who are involved in clinical research — regulatory practitioners, clinical trial designers, trial implementers, management of sponsor companies, or contract research organizations — to be fully aware of and committed to these principles because, if for no other reason, it is the right thing to do.

7.4 IRBs and IECs

As noted above, the term Institutional Review Board (IRB) is used in the United States and Institutional Ethical Committee (IEC) is the term used in other countries to designate the group responsible for safeguarding the rights, safety and well-being of subjects participating in clinical research over which the group has jurisdiction. They serve essentially the same functions regarding clinical trials and will be considered together here.

We will address responsibilities, their composition, functions, operations, types of procedures that they put in place, records and record-keeping, and we will cover items related to advertising. Regulations covering activities of IRBs in the US are contained in 21 CFR 56 (http://www.tsghirb.ndmctsgh.edu.tw/doc/cfr_21-56.pdf). The ICH Guidance is ICH Topic E6 (R1), Guideline for Good Clinical Practice (http://www.ich.org/LOB/media/MEDIA482.pdf).

7.4.1 Responsibilities

The purpose of the IRB is to review the research and to determine if the rights and the welfare of human subjects involved in the proposed research are adequately protected. By institutional policy, even studies that may otherwise be exempt from United States Federal regulations may require review and / or verification of exempt status. So "institution" in this case means the institution where the study is being reviewed.

Documents provided to peer review committees and IRBs by investigators must contain enough information to allow valid judgments about the science and the ethics of the research.

The IRB has the authority to approve, to disapprove or to require modification in all research activities. They also can require changes in previously-approved human subjects research studies, particularly those for which they've received new information.

One of the important considerations is the content and focus of the review. Many institutions will apply this particular set of responsibilities somewhat differently. There are institutions, for example where I did my cardiology training years ago — Mt. Sinai in New York City — where the IRB was very detailed and conservative in application of its review and oversight function. They spent a lot of time looking at all of the elements of the research protocol, particularly at the design of the proposed study. At the time (in the 1980s), this approach was somewhat unusual. In current times, more IRBs, including central IRBs, will at least address the design elements. Other IRBs may spend more time looking at inclusion / exclusion criteria, laboratory procedures (particularly the amount of blood withdrawn), randomization procedures, etc., as part of their review of clinical research.

Some of the documents required by IRBs and ethical committees include, obviously, the protocol and its amendments. Some IRBs will require the protocol and its amendments be summarized separately from the summary provided in the protocol.

Most protocols with amendments will contain a revisions page so changes from the original can be easily identified. Some sponsor companies don't provide the detailed revisions page; they simply will identify on the front cover that this is a new revision, dated such and such. Having been part of an IRB in the past, I can tell you it is a whole lot better if the reviewer knows exactly where the changes to the protocol have been. The reviewer is then better able to assess those changes in terms of their overall importance for this particular protocol and study and, in turn, can better assess benefits and risks for the patient. I would suggest being kind to your friendly IRB reviewers and put in the changes section.

In addition, the patient consent forms need to be provided. A bit later in this chapter (Section 7.5), we will take a look at the required elements of the consent. As the IRB reviews the consent, they will ensure that all the required elements are contained within the proposed consent.

Whatever recruitment procedures will be employed for the study must be identified to the IRB. Those procedures may include: "We are going to a physician's office to check with him or her about how many patients they have had in the past who might be called to come in for screening to see whether they qualify." or "We will recruit patients from today going forward only." The former method comes under the purview of HIPAA which we will address later in this chapter (Section 7.7).

On the other hand, if we are going to apply other types of recruitment procedures, such as newspaper ads, television ads, radio ads, the actual text that we are going to use as part of that ad must be submitted to the IRB for review and approval.

In addition, whatever other information will be provided to the study subjects must also be provided to the IRB for their review.

We discussed the Investigator's Brochure (IB) in Section 5.2. The version submitted needs to be the latest available. Typically, a company will revise the IB on an annual basis and include non-clinical, pre-clinical, clinical, etc., information that has accumulated over the past year. At the time of IRB submissions, depending on information available and the recency of the latest version, the IB might be updated more frequently. If there is very recent information that materially impacts the potential benefit:risk assessment, an addendum to the IB might be a reasonable vehicle through which to provide that information to the IRB. Providing this information enables the sponsor to put the new data within the context of other available information.

Subject compensation is provided to subjects for their participation in the study. Referring back to the discussion earlier in this chapter, the Belmont Report specifically addresses the issue of "undue influence". We could say, for example, to the patient: "Hey, participate in the study. Take this antihypertensive drug (or the placebo) every day for the next month and we will give you $5,000." That would seem to be, surely to me, an inducement out of proportion to the risk the patient would experience as a result of participating in this particular trial. That type of compensation would need to be identified to the IRB. It's fairly common to provide transportation cost reimbursement to patients, particularly in urban centers where it may be more difficult for them to get around. Patients who may, for example, have congestive heart failure — the target subject patients for the study — may have difficulty walking. In that case, for example, the sponsor could provide the patients some form of transportation to get from their home to the hospital, etc. I would generally consider that financial "incentive" to be reasonable, but it, nevertheless, needs to be identified to the IRB.

All the qualifications of the investigators need to be provided to the IRB or to the ethical committee. No matter how well known the investigator is within the institution (indeed the individual might be a member of the IRB), adequate documentation of that individual's competence, expertise and training that enables them to be a qualified investigator for this particular study needs to be submitted to the IRB in the initial data package. Review of the investigator's qualification should also go beyond the review of their CV. Certainly, there are multiple examples of physicians who are well qualified by review of their CVs with regard to training and experience, but who turn out to be essentially incompetent when it comes to organizing and conducting clinical research. Investigators, on the basis of incompetence or active fraud, have been banned from participating in clinical trials. Thus, it is also within their purview for the IRB to make an assessment of an investigator's trial competence and judgment. (While I recognize that, within an individual institution, such an assessment is fraught with political hazard, it is, nevertheless, a reasonable concern on the part of the IRB.)

There needs to be some mechanism in place for the IRB to provide on-going review of the trial. It's required that the IRB review the trial at least on an annual basis. The IRB, in considering the therapy and the disease being addressed, might request a review within a shorter time frame — that is certainly within the scope of their responsibility.

Finally, the IRB has specific responsibility, as noted in the discussion of the Belmont Report, regarding the protection of vulnerable subjects, e.g., minors or those who might be incapacitated, to make sure that their participation in the study is as voluntary as is possible.

7.4.2 Composition

The composition of the IRB needs to include "a reasonable number" of members, but at least five. The precise composition obviously depends on the institution, location, etc., but the number needs to be large enough to ensure quota at meetings and to ensure that the workload is not burdensome for members, yet small enough to ensure adequate discussion regarding potential risks of studies.

One member needs to be a non-scientist. Typically, such individuals come from the local community. Many times, they are somewhat older individuals who might be retired and are able to participate in this type of activity on an on-going basis. On the IRB on which I participated, we had a person who was retired, in his late 50s. Initially, his review was fairly straightforward. I might even label it "simple" from the standpoint of understanding the details of the protocol and risks and benefits of the prospective therapy. However, he brought a sound sense of judgment, a natural sense of curiosity, and, over time, an increasing level of sophistication and understanding that made him a very valuable member of the Board. Indeed, he was eventually able to go toe-to-toe with the scientists and physicians regarding the rigor of trial design. In particular, he didn't accept something simply because "that is the way it has always been done."

At least one independent person, without a prior relationship with the institution, must also be included on the IRB roster. On the IRB I mentioned above, the non-scientist was also the independent person. Where the non-scientist or independent person(s) come(s) from specifically is not defined in the regulation; the point is to make sure the review of protocols and issues by the IRB is done in as objective a framework as possible.

A list of the members and their qualifications needs to be maintained by the IRB / ethical committee. This is a document that the sponsor needs to obtain prior to study commencement. The list will include all current members of the IRB, their terms of membership, their qualifications, and the functions they are performing on this particular board. The list would designate who, for example, the chairman, the vice chairman, the non-scientist person, and the independent participant are.

7.4.3 Operation

The IRB needs to have a set of standard operating procedures and keep written records. Some of these things seem very obvious, but the sponsor needs to review the SOPs and at least document the presence of written records prior to the start of a trial.

The IRB must have announced, regularly scheduled meetings too for review and decision making. The exact schedule is, of course, up to the institution taking into account IRB membership and the volume of clinical research. The sponsor wants to know the timing of meetings (and the submission dates for getting on the meeting agenda) so it can optimally time its submissions.

There may be special meetings required for approvals of certain protocols. It is likely that, as part of the administrative standard operating procedures, there may be procedures in place that would enable a subset of the board to approve a protocol that had already undergone review and had minor textual changes or changes in the consent and so forth. Thus, there could be expedited reviews.

For each meeting in which decisions are made and votes are taken, there must be a quorum present. Only members who participate should vote. So, in other words, if I happen to be stuck at a hospital lounge and I can't quite make it to the meeting, but I know that my friend Joe's protocol is going to be reviewed, I can't send in my vote in absentia by telephone or by letter and say, "Hey, I think Joe's protocol ought to be approved." If you are not there, you can't vote on that particular protocol.

It is also important to note that, even if an investigator is a formal member of the IRB, he or she may not participate in any of the deliberations regarding a protocol in which they would have a role. They may respond to questions or requests for information, but may not, otherwise participate.

7.4.4 Procedures

The specified procedures should cover the initial and ongoing review of trials, what the procedures are for notification regarding protocol review, approval, clarification and change. In addition, the IRB will also publish procedures identifying requirements such as trial admission, consent approval, reporting of deviations, trials risks, adverse drug response or adverse events.

7.4.5 Records

Records need to be retained for at least three years after the conclusion of an individual trial. The records may be available to investigators, to sponsors and to various regulatory authorities. The time periods are specified in the regulations. The informed consent forms patients review and sign must identify that records are subject to review by investigators, sponsors and regulators.

7.4.6 Advertising

Advertising for recruiting subjects into the study needs to be reviewed by the IRB. Both the content and the mode of communication need to be reviewed. Whether it's a video ad for TV, an audio ad for radio, or a print ad, the IRB should review it to ensure that the information is not misleading to potential subjects. It also ensures that no claims are made, either explicitly or implicitly, that the drug or the device under investigation is safe or effective. It also needs to ensure that no claims are made that the drug or device is equivalent or superior to any other drug or device.

The advertising needs to include the purpose of the research (generally including a minimal set of inclusion criteria) and information regarding contact with the clinical site.

Generally, such advertising points to a screening location — telephone or web site. At this point, the more specific inclusion and exclusion criteria are presented as well as an indication of the potential compensation for the trial. Once the potential subject passes this screening stage, they are typically interviewed again, in person or on the telephone, prior to actual study consent. The process is designed to weed out those not likely to qualify as quickly and as cheaply as possible.

This whole scheme, including a description of the process, the content of the initial ads, and the screening scripts used prior to the subject's visit to the trial site are all reviewed by the IRB.

7.5 Informed Consent Forms (ICFs)

7.5.1 Essential Elements

For additional discussion of the Informed Consent Form (ICF), please see Section 4.4.8.

Essential elements of the ICF include a statement that the trial involves research, a statement of the purpose of the study, an approximation of the number of subjects involved, a description of the available treatment possibilities and the randomization probabilities, and a description of the trial procedures, including potential risks of those procedures. Subject responsibilities also need to be delineated, including the duration of the subject's participation.

Risks or inconveniences, as well as expected or potential benefits, need to be specified. A description of alternative procedures or treatments, if any, should be included. Payments to subjects for participation are specified as well as what care / compensation are provided in the case of adverse events.

The Consent Form must specify CLEARLY that participation is voluntary and the subject may withdraw, without prejudice, at any time. Also, general circumstances for subject or study termination must be listed.

As noted above, it is important to state in the Consent Form that records will be maintained within a confidential environment and that the records will be accessible to representatives of the sponsor, the IRB, and regulatory authorities.

The Consent Form will also identify and specify the contact procedures in case of emergency.

7.5.2 Common Mistakes

Several mistakes are commonly noted in dealing with informed consent forms (ICFs). Not using the required format is a common mistake. Submitting a consent form from another institution or a study sponsor is a common mistake. It does happen that somebody at the research center, the clinical coordinator, will pull off the shelf a consent form that happens to be used by World Famous Pharmaceutical Company 2 instead of World Famous Pharmaceutical Company 1 or it could be that World Famous Pharmaceutical Company 1's consent is used for a study that actually is fairly similar to the study that your company, Super Small Biotech, is conducting, and their consent is used by mistake in your study or vice versa.

The language in which the consent is written is required to be at approximately the 7th or 8th grade reading level. Many consent forms are written at too high a level and also include too much technical jargon. Of course, the ICF written in English might present challenges for an individual whose primary language is not English. Occasionally, the forms are written in such poor English than an English-speaking reader can't understand them either.

Another common mistake is that procedures are not clearly explained. "You will undergo a phlebotomy, and as a result of the phlebotomy, you might experience syncope." The average reader won't know what a phlebotomy is (and is likely to not understand syncope either).

Another mistake is not including all the required elements identified in Section 7.5.1. A patient's bill of rights is a set of rights that the patient has as a result of participating in a clinical trial; it needs to be attached to the informed consent.

Finally, a not uncommon mistake is use of an earlier consent form that has been replaced by a newer one or, alternatively, use of a more recent consent form prior to its formal approval by an IRB, or use of a consent form that did not include changes specified by the IRB. The FDA will audit these consent forms ultimately, so it is crucial that the correct forms be used.

7.5.3 Challenges in ICF Preparation

In preparing informed consent forms, there are some challenges. For example, avoid overly optimistic language.

For long and complicated studies, be careful to stress the duration of participation — it requires a commitment on the part of the subject — and that they need to keep their copy of the consent form for the duration of that particular study. While a copy is kept at the investigative site, it is important they keep the ICF also — it has study contact information and other details pertaining to the study included on the form.

The original consent form kept in the investigator's records needs to be signed and dated BEFORE any research takes place. REALLY!! Although mentioned previously, it really does bear repeating again.

We have previously mentioned language as an issue. In the US, in California, Texas and Florida, Spanish is a prevalent language and indeed many potential subjects speak little English, so a consent form for those individuals would, appropriately, be in Spanish. As a reminder, many individuals, English-speaking or Spanish-speaking, do not know how to read. Accordingly, for their participation in a clinical study, consent has to be handled in other ways to ensure a fully-informed participation. Recently, I conducted a study in the Bay Area, in Oakland, California, and a large number of the subjects came from a local Chinese community. The Consent Form was translated into Mandarin (and the translation validated) to ensure that participating individuals would have the choice of reading the consent form in either English or Chinese.

As trials are conducted outside the United States, the language used needs to be appropriate to the particular country and also understandable by the local population. I was involved with a study recently that was conducted in Eastern Europe. The questionnaires and the Consent Forms had to be translated into Hungarian, Serbian, Russian, Ukrainian, Croatian, etc. Each also had to be back-translated and officially validated. Both the English version and the local language version were reviewed by the relevant Ethical Committee.

Competency of the subject is also important to establish prior to reviewing the document and receiving a signature. We have discussed earlier obtaining consent in children. An interesting consideration would be a study conducted in patients with Alzheimer's disease. How does one go about getting the consent for that study? In that case, it would be appropriate to establish the protocols for providing information and obtaining consent from the patients and their family or care givers prior to starting the study.

7.5.4 Audits

The sponsor should also plan, in addition to standard monitoring, to audit the ICFs used in the trial. (A regulatory authority auditing the site will do the same thing.) One issue is timing. The principal investigator's signature should never pre-date the patient's signature. The patient's signature should also be dated and timed prior to any study-specific procedure.

Another issue might be the version of the consent. Not infrequently, an ICF is revised during the study (for a multitude of reasons). At the very least, the version of the ICF being signed by the subject should be the version in effect at that point in time. If the ICF changes during a study, it is common to find that subjects are being provided, and signing, the wrong version.

An additional consideration concerns individuals who, for example, signed Version 3 (the appropriate consent) at the time of entry into the study. However, during their continued participation (perhaps study visits stretch over a year), a new ICF, Version 4, becomes operative. Most regulatory professionals — not all, I might add — recommend having the patient sign the new version as well (particularly if the changes are "substantive" with regard to the conduct of the study or safety).

The FDA focuses on consents during audits, so the sponsor and the site should also.

7.5.5 On-Going Review

As noted above, IRBs are obligated to conduct periodic reviews of on-going research. Often, this oversight is missing.

Although IRBs are to review each study on an annual basis, at least, it is their obligation to review studies more frequently if there are new data regarding the compound or other information that would potentially change the risk:benefit assessment.

For example, the sponsor's initial study, a Phase I study in normal subjects, is the first trial in humans. At this point, only a set of non-clinical data are available to suggest what the risk profile might be; no other human experience is available. The IRB is certainly well within its rights to say, "Let's take a look at the first five people, 10 people, first three months, six months, etc., and let's re-review this to see where we are." Similar logic might also be applied to other Phase 1 and early Phase 2 studies, resulting in more frequent IRB review.

Information contained in the Investigator's Brochure suggesting the possibility of a specific risk, a change in principal investigator for the study, or use of an investigator with prior compliance problems (hopefully, that person wouldn't have been chosen in the first place) might be additional reasons for more frequent review.

As part of the continuing review, the IRB must determine whether the risk:benefit ratio remains acceptable. Information should be available regarding the local site as well as communications from the sponsor regarding events or issues at other trial sites, data obtained from other clinical studies, and data that might derive from newly-reported non-clinical studies that might have an impact on perceived safety in the trial.

7.6 Compliance

The Common Rule, contained in 45 CFR 46, deals with compliance. It was initially adopted in 1991 and was most recently revised in 1999. It has been adopted by 17 different US Federal agencies and it covers multiple aspects of clinical research. Subpart A is a basic policy for covering human research subjects. Subpart B specifically deals with research and fetuses, pregnant women and *in vitro* fertilization. Subpart C looks at prisoners and Subpart D looks at children as subjects in research. To stay current in areas of compliance in clinical trials, review information on these two web sites:

- FDA Office of Compliance
  Monitors sponsors, investigators / CROs / IRBs and IECs
  http://www.fda.gov/Drugs/GuidanceComplianceRegulatory Information/default.htm

- Clinical Investigator List
  Listing of bad (and good) investigators
  http://www.accessdata.fda.gov/scripts/cder/cliil/

As an illustration of the types of problems encountered by the US FDA, Table 7-2 lists examples of letters issued by the FDA to investigators and investigative sites (including IRBs) in 2007. If you pick another year, you will find similar numbers. It seems amazing that these issues keep cropping up year after year.

Table 7-2. FDA Compliance - Warning Letters (2007)

14 Letters issued to investigators / investigative sites in 2007	
Protocol non-compliance	86%
Informed consent non-compliance	79%
Poor record keeping	57%
Poor AE reporting	43%
Failure to protect subjects	29%
Inadequate documentation of IRB approval	29%

7.7 HIPAA

The Health Insurance Portability and Accountability Act (otherwise known as HIPAA) was enacted by Congress in 1996. Title I was designed to protect workers as they moved from job to job — or if they were to lose their job — by enabling the continuing coverage of health insurance in those transitions. Title II was designed to address national standards for privacy in electronic health care transactions.

The rule went into effect in 2003 (or 2004 for small health plans) and covered "certain individually-identifiable health information by establishing conditions for its use and disclosure".

How does it impact clinical research? Well, it all depends . . . It depends on the location of the work environment and the type of information desired.

We have, in my opinion, three overlapping sets of rules and regulations: the Privacy Rule, the Common Rule, and FDA Regulations. The Privacy Rule does not replace or modify the Common Rule or FDA Regulations. The Privacy Rule, again in my opinion, is an additive layer to the privacy protections of the Common Rule and FDA Regulations.

HIPAA applies to something called "covered entities" regardless of where their funding or income comes from. It contains standards for the de-identification of health information. Importantly (I know all of you have signed multiple forms over the past few years acknowledging that HIPAA regulations are in effect at your health care provider's office), it requires authorization of disclosure of that information. The rule also applies to decedents' information.

7.7.1 Covered Entity

What is a "covered entity"? By legislation, it is "a health care provider who transmits health information electronically in connection with a transaction for which the Secretary (of HHS) has adopted standards." Fundamentally, it means that any health plan, health care provider, or health care clearinghouse (anyone doing claims processing, etc., on behalf of health care providers) that provides health care services and conducts any transactions electronically falls into the definition. (There are some nuances, but not really important for our purposes.)

7.7.2 Protected Health Information

The next key component to the rule as it applies to clinical research is the definition of what is protected health information (PHI). PHI is information that is transmitted or maintained in any form (paper, oral, electronic, forms, web-based, etc.). It, however, does not include de-identified health information or biological tissue; there are also certain other exceptions (e.g., some employment records or education records).

A formulaic way of looking at it is this:

Protected Health Information (PHI) = Covered Entity + Health information + Identifier

Essentially, this means that any information that can be used to identify the owner (i.e., patient) is protected. For the industry, this means, essentially, that all subjects of clinical research must sign a release that specifically addresses HIPAA regulations if they are to be included in the study. There are certain limited circumstances in which individual authorization is not required. It is always best to check the latest information. This is available at this web site: http://privacyruleandresearch.nih.gov/.

7.7.3 Research Authorization

What is sufficient authorization? It must be for a specific research study (rather than being a blanket release for any research). For example, while authorization for use or disclosure of PHI to create or maintain a repository or database for a specified study is permissible, authorization for future, unspecified research is NOT permitted — even if a patient's signature is attached to the statement. It is important to understand that authorization is not identical to informed consent, although it can be combined with the informed consent process and signature.

Much as the informed consent itself, HIPAA authorization must contain certain "core elements" and a set of "required statements". As for the informed consent, a copy must be provided to the individual. The expiration date (or non-expiration) must be clearly stated on the form as the patient signs it.

Table 7-3 lists Core Elements required for an authorization for use or disclosure of PHI.

Table 7-3. Core Elements in PHI Authorization

Core Elements
Description of PHI to be used or disclosed
Person(s) authorized to make the requested use or disclosure
Person(s) to whom the covered entity may disclose PHI
Each purpose for the use or disclosure
Expiration date or event (e.g., "end of the research study" or "none")
Participant Signature and Date

Table 7-4 lists a set of statements to be included in the authorization.

Table 7-4. Statements Used in PHI Authorization

Right to revoke the Authorization (plus exceptions) and the process for revocation
Ability or inability to condition the treatment, payment, or enrollment in the research or eligibility for benefits based on acceptance of Authorization
The PHI may no longer be protected by the Privacy Rule once it is disclosed by the Covered Entity in the process of the research.

While the information itself is really self-evident, the process of obtaining the authorization does impose an additional burden on an already detailed ICF process.

7.7.4 Impact of HIPAA

The Office of Civil Rights (OCR) is responsible HIPAA-related complaints.

A recent article estimated the burden HIPAA imposes on clinical research. At least 2 million people per year sign authorizations for clinical research. Currently, the complaint rate is about 1 per 600,000 for clinical research. This compares to a figure ten times higher (e.g., 1 per 60,000) for general medical care. In fact, only 17 complaints over a 5-year period were related to clinical research. The article estimates that at least $10 million per year is spent obtaining the HIPAA authorizations. (Goldfarb, 2008)

7.8 Ethics and Compliance Resources

Table 7-5 identifies various regulations and forms that apply to ethics and compliance.

Table 7-5. Ethics and Compliance Regulations

Regulation	Reference
Privacy and Protection of Subjects	
Electronic Records; Electronic Signatures	21 CFR, Part 11
Human Subject Protection (Informed Consent)	21 CFR, Part 50
Additional Safeguards for Children in Clinical Investigations of FDA Regulated Products (Interim Rule)	21 CFR, Part 50, Subpart D
Institutional Review Boards	21 CFR, Part 56
Financial Disclosure by Investigators	
Financial Disclosure by Clinical Investigators	21 CFR, Part 54
Drugs and Biologics	
Investigational New Drug (IND) Applications	21 CFR, Part 312
Form 1571 (IND Application)	
Form 1572 (Statement of Investigator)	
Applications for FDA Approval to Market a New Drug (NDA)	21 CFR, Part 314
Application for FDA Approval of a Biologic License	21 CFR, Part 601
Devices	
Investigational Device Exemptions (IDE)	21 CFR, Part 812
Pre-Market Approval of Medical Devices (PMA)	21 CFR, Part 814

References

Beecher HK. Ethics and Clinical Research. New Engl J Med (1966) 274:1354-60

Goldfarb NM. HIPAA Complaints in Clinical Research. J Clin Res Best Practices (2008) 4(2):1-3

Milgram, S. Behavioral Study of Obedience. J Abn and Soc Psych. (1963) 67:371-378

Shuster E. Fifty Years Later: The Significance of the Nuremberg Code. New Engl J Med (1997) 327:1436-1440

Part II

Executing the Plan: Do the Right Things Right

Chapter 8

Project Management

This chapter will initiate our discussion of the operational aspects of clinical trials. First, we will look at an overview of project management. Subsequent chapters in this section will address issues relating to clinical trial monitoring and site management.

The key processes associated with project management in general work very well for pharmaceutical projects in particular. (See Table 8-1.) Large books have been written covering various aspects of project management. It would be a bit presumptuous to assume I can cover all the key aspects in a single chapter. However, I will highlight those items I think are crucial for improving the likelihood of success within the clinical development area.

8.1 Project Definition and Project Constraints

To get started, a clear direction and an understanding of the steps required to get from here to there are needed. Clear direction is provided by the package insert or the Target Product Profile (Chapter 3). Identifying these elements — as they relate to demonstrating efficacy while describing safety — enables translation into the key steps to incorporate into the Project Plan.

However, all projects are constrained in several dimensions: time, money, people resources and quality. By virtue of those constraints, there will always be trade-offs.

Table 8-1. Key Processes in Project Management

Identification of key stakeholders
Definition of project team members
Determination of the ultimate objectives of the project (project requirements and scope)
Breaking project deliverables into discrete components (work breakdown structure)
Definition of the activities associated with the objectives
Determination of the sequence of activities required, resources needed, time required
Development of a schedule
Determine costs and budget
Define plan quality
Develop human resource plan
Describe communication plan
Determine project risks and plan risk responses
Execute the plan
Monitor the plan
Modify the plan (if needed)

The start of planning a development program is always a definition of the scope of the project. By using the Target Product Profile, we can determine the scope of the activities required to move our project through to registration. As a reminder, I believe the plan should be prepared as though you are carrying the project to registration, even if the corporate goals would suggest that an earlier value event — partnering, licensing, M&A —would occur before that point. In any case, it is important to determine what the project should achieve.

At the beginning of the planning process, it is very important to identify the full range of stakeholders and their needs. The plan must address those needs. There is certainly a distinct possibility that the needs of various shareholders might conflict. In that case, it is best to have an open discussion at the beginning of the project and attempt to resolve differences. I won't pretend the resolution process will go smoothly, but I really believe it is important to get everyone to buy into the goal and the project at the beginning. Trust me, it is better that way.

Any project manager will tell you that a key element at the beginning of a project is to define the standards to which it will operate. Perhaps a slightly different example might illustrate the point. When filing my first NDA, I was the designated reader, i.e., I got to read each and every section of the NDA from cover to cover. By actual count, I read some sections more than 10 times. On the 11th time, I could still find typographical mistakes (not just style differences) in sections I had read previously. However, by this point, they were so few and far between that their significance was minimal. Keeping in mind that SmithKline hadn't filed an NDA in a while and they were under a consent decree with the Agency for a prior problem, there were folks at the time who "demanded" (that is the right word) that the NDA not leave the building until it was perfect. Fortunately, clearer heads prevailed — in reality, there is no such thing as a perfect document, particularly one stretching over thousands of pages. What we did know, as we delivered it to the FDA, was that all sections made sense and all references to other sections had been verified. In fact, we received comments / questions from all three major section reviewers (clinical, pharmacology / toxicology, and CMC) within three weeks of the time it was delivered. That was BEFORE it was officially filed! That meant it was readable enough that people got right through it.

The point of the story is that perfection is unattainable. Life — no matter what part we are talking about — is messy. As a project is defined, decide on the level that really is "good enough". However, once the standard is set — do not compromise!

By definition, a project that needs management is a project that is constrained. So, at the beginning of the development planning process, identify the constraints. The four biggies, as noted above, are time, money, resources (people), and quality.

Let me spend a moment on money. If the sponsor's organization is a smaller enterprise and relies on venture capital support, it is a very useful exercise at this point to identify how much money is available and what the run time (i.e., until the money runs out) is. Run different scenarios regarding different strategies for development. I would suggest the plan developed be integrated with points in time at which money will be needed to be brought into the company. It is very important that useful, supportive data be available in advance of those "dry well" times to enable the company to raise money at the best price possible. Even if those funding points are "predetermined" to be licensing, partnering or M&A points, plan for the unexpected — what happens if those licensing, partnering or M&A deals fall through or the market just gets ugly?

Assess the organizational support. Part of that is an assessment of emotional and psychological support and commitment for the project. The other part is resource identification — we have discussed previously about filling holes the organization might have with regard to certain competencies. The latter analysis includes counting bodies, but, more importantly, includes a determination of levels of expertise in the required areas. If the resource or expertise is not available within the company, the decision becomes a "make or buy" decision — do I hire somebody ("make") or do I rent a consultant ("buy") for the expertise? The answer to that question is always individual to the specific company, the specific requirements, and the resources available.

In addition, it is crucial to identify the potential risks to the plan that we are working on and to formally develop contingencies. Recalling Metternich's famous "Plan B" back up strategy, I believe it is absolutely essential to identify, as much as possible in advance and as much as possible in writing, what the contingency plans will be in case something goes wrong. If IT (whatever "it" is) is identified as being critical to the project, IT deserves a contingency plan. There are other risks that should be identified, but perhaps we don't need to fully mobilize a contingency plan today; at least identify it. But, for that first set of risks, the ones that we think could significantly derail our project, we need to identify in advance and, for them, develop full-scale contingencies.

Working on my first NDA, in prehistoric times, required a focus on paper and making multiple copies. Clearly, one item was to make sure we had enough paper on hand. In addition, we also had to make sure we had a contingency in case the photocopying machine broke down. Part of the contingency was a repair contingency and part was a commitment on the part of a repair company that they would literally bring in a new machine if we needed to have that done. A critical point has been identified; a contingency plan was put in place. While a simple example, that's the kind of contingency that must be planned for and must be resourced at the beginning of the project.

8.2 Set Goals

In addition, as the project is defined, one of the key challenges is setting goals. Goals must be clearly defined — it must be obvious when they are achieved or when they aren't. That being said, some other points are equally critical — after all, ultimate success is desired. The goals must be achievable, they must be realistic and they must be consistent with the project scope.

Of course, there might be a hair-pulling day when there is a realization the company's CEO just committed to some analyst or at some road show or to a VC that the answer to this project will be available on such and such a day! The project manager knows the schedule so well he or she can quote it in their sleep. Just to be sure, though, the project manager re-checks. Right! Not a snowball's chance in a very warm environment! The project manager might need that come-to-Jesus moment with the CEO. On the other hand, hopefully he or she is educable and problems like that can be avoided in the future.

Nevertheless, the project — while a stretch — is achievable. Assuming that is true, people can be incredibly motivated and perform magnificent things. But, if everyone knows it isn't possible, no matter what, those same people can be extremely demotivated.

8.3 Identify Milestones

Break the project into phases and milestones and make each of those milestones meaningful. It is optimal to have milestones connected with decision points: "We proceed." or "We don't proceed." It is generally best if the "call option strategy" we discussed previously (Chapter 1) matched these milestones.

Sometimes milestones are built in for emotional and motivational factors and that's okay. Understand what that particular milestone really accomplishes in the plan. Intermediate milestones are also useful as progress markers and motivators.

8.4 Critical Path and Resource Leveling

Identify each of the tasks to be completed within the phase of the milestone. If tools such as Microsoft Project or other project planning software are used to map and monitor the project, it is absolutely crucial to put *all* of the information in the program, including all of the component sub-tasks that are required for achieving that ultimate goal. Each of the major software programs enables input of the full **work-breakdown structure** (all the items that go into any particular task), and to identify the resources required as well as the duration for each of those tasks and sub-tasks.

Do not compromise. Put all the information into the project management software. It does require a lot of time up front, but having the information, including all the relevant dependencies, in the program enables an assessment of resources and permits resource leveling. Furthermore, the ability to do "what-if" scenarios is really crucial. As part of the process, the critical path is identified.

The **critical path** is the set of activities that, if anything changes, the time line for achieving the success of the project changes. You need to determine that critical path and manage to that critical path.

Make sure that the project operates with honest resource leveling. For example, a person is doing activity A and they are assigned for 35 hours a week. "Hey, no problem!" They've got five spare hours. Piece of cake. Well, actually, in the environment in which they are working, that person has 15 spare hours. We know they can achieve that particular activity during the time allotted. On the other hand, we now take a look at that set of activities and realize the person is also working on another component of the project and, by the way, they are assigned 35 hours for that one too. PROBLEM! Simple math, of course, tells us resourcing needs to be adjusted.

That kind of resource understanding is crucial to success.

8.5 Refine Constantly — Test Assumptions, Never Assume

Once the project information is input into the software, keep looking at it, looking at it, looking at it. Refine it, refine it, refine it.

Constantly test all of your assumptions and always evaluate every change that occurred for an impact on the schedule or the resources.

Manage the project prospectively. By that, I mean someone has to be looking down the road — what is scheduled for next week, two weeks from now, a month from now? Someone must be checking to ensure that those future activities are on track for completion. If, for some reason, it appears that the future activity is at risk, the antecedent tasks can be reviewed and the problem identified before it becomes a big problem. Resources can be applied or other changes made in order to preserve the time line.

Time is always an issue. Once the processes are identified and the contingencies specified, review the processes. Re-engineer them to minimize elapsed time. Identify non-value-added activities and eliminate them. Try to conduct as many value-added activities in parallel (always being mindful of available resources). Eliminate dead time as much as possible. Identify alternative resources or additional resources and reduce capacity-driven bottlenecks.

Oh, by the way, do this while keeping costs down and quality up!!

8.6 Project Teams — Who?

Different organizations label things differently. Sometimes there are project teams and sometimes there are product development teams.

Occasionally, both are present. At the least, there needs to be clear responsibility for the creation and management of the plan to get the product through the development process.

We have previously identified the functional expertise required for development of the plan. Essentially, that same group of individuals / functional expertises needs to be included in the management of the project. The set of core expertise required includes (at least): drug discovery, drug formulation, pathology and toxicology, pre-clinical research, clinical research and development, regulatory affairs, technical operations (including manufacturing), and marketing.

The resourcing question remains: Are the resources available inside the company or do we need to go outside to ensure adequate expertise?

8.7 Project Team — How?

The simplest way of approaching the management of the project is to mirror the development plan. Base activities on the key milestones, such as drug candidate definition (or declaration), submission for clinical trial initiation, submission for marketing approval, and marketing launch (at least as major milestones).

Skills are important. Probably the key skill, to my way of looking at project management, is prospective management. It is potentially as much an attitude as it is a skill. But prospective management applies to everything: scope, time, cost, quality, human resources, communications, the integrated project. As noted in Section 8.5, this is a process of looking at least 2-3 steps ahead and asking if the requisite predecessor tasks are on track, and if the resources needed to accomplish the tasks are in place, AND to work through potential contingency plans in cases something doesn't work as planned.

Tools are also important. There are lots of tools available — lots of very powerful software programs that enable one to manage particular projects. They all provide multiple means of portraying the information — lists, Gantt charts, PERT charts, etc.

What the tools enable you, the project leader / the project member, to do is to identify and to manage your own resources and milestones. The tools, once you apply them and know how to use them, are in fact very helpful and I would say in some cases, even magical in terms of enabling the understanding of relationships. The ability to do resource leveling is a particularly valuable aspect of the software.

Figure 8-1 is an extremely simplified version of a clinical study time line illustrating the Critical Path concept and time relationship of various activities in the project plan.

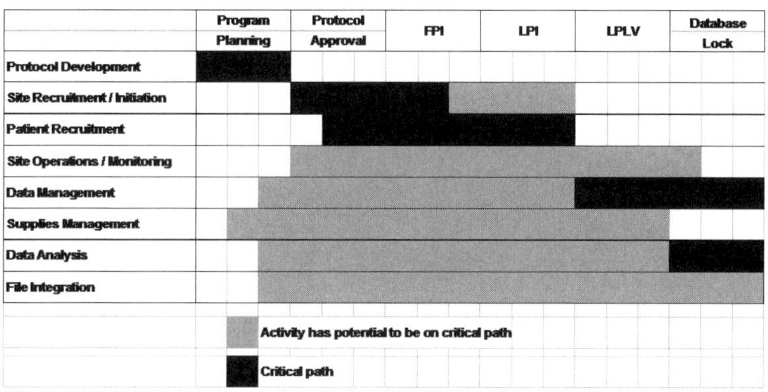

Figure 8-1. Example of Clinical Trial Time Line

FPI = first patient in; LPI = last patient in; LPLV = last patient last visit; CSR = clinical study report.

One minor — Okay, really major — consideration, however, is that THE TOOLS (SOFTWARE) DON'T MANAGE THE PROJECT, PEOPLE DO!!

The value of the tools lies in several key aspects. First, to enter the information in a logical and related fashion, one is forced to analyze all the activities required for the project and to understand the respective dependencies. One aspect, not to be lost, is the value of simply being able to represent the information in a visual fashion. There is also — assuming the information is input correctly and appropriately — a forced honesty.

But the resulting power is an incredible ability to "what if". This is the capacity to look at the project plan and say, "If this goes sooner . . ." or "If this goes later, what happens to the overall project time?" The ability to "what if" is crucial in terms of planning for contingencies and affording the team the best opportunity for success at the end of the process.

Chapter 9

Clinical Operations

Figure 9-1 identifies, in a highly schematized fashion, the core activities associated with the conduct of a clinical trial. Core activities of clinical operations broadly fall into these categories:

- Definition and implementation of clinical strategies
- Preparation of clinical project time lines and resources plans
- Identification of key opinion leaders
- Qualification and selection of investigators and investigational sites
- Identification of study-specific patient recruitment strategies and programs
- Negotiation of contracts / fees with investigators and sites
- Preparation, assembly and maintenance of study documentation, including protocols, investigator's brochure, and case report forms, binders, investigational products, clinical supplies, etc.
- Hands-on management of sites, including monitoring visits, enrollment, queries and query resolution, verification of data with source documents
- Reporting and tracking of adverse events, including serious adverse events
- Preparation, shipping, storing, distributing, and accounting for all clinical study material

This chapter will focus on all activities having to do with the site and site management. Subsequent chapters will deal with issues related to data management and safety.

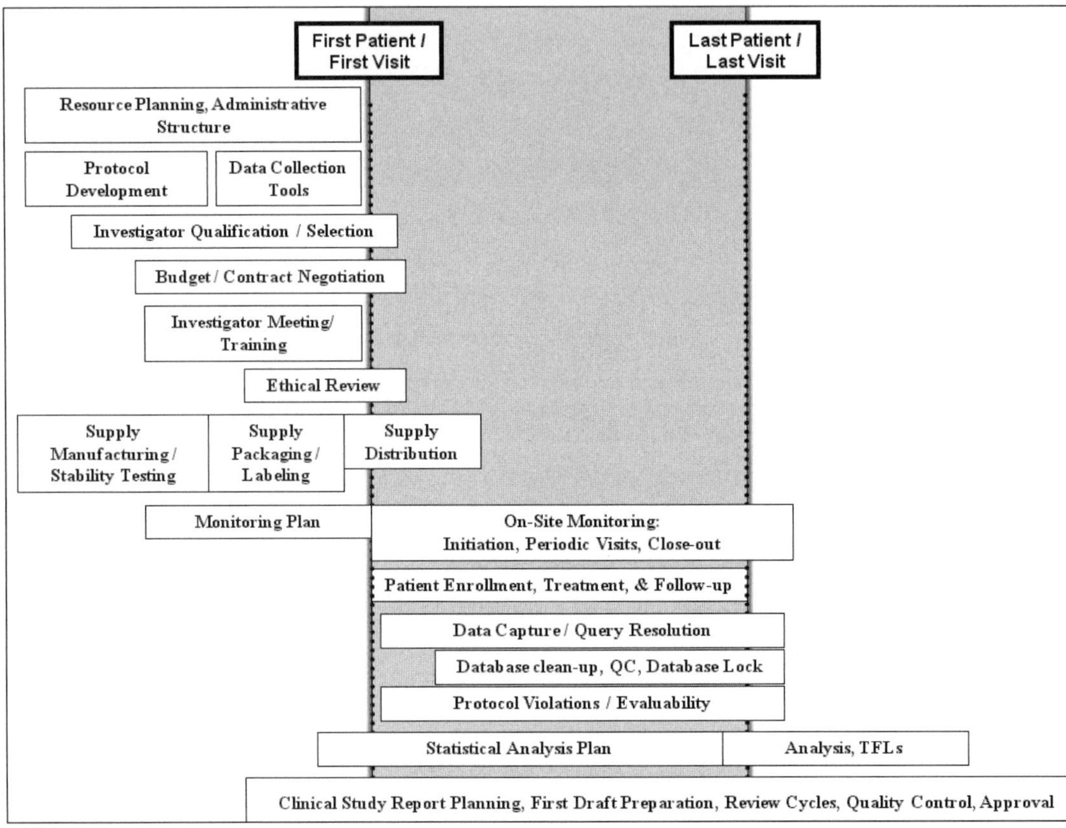

Figure 9-1. Clinical Trial Activities

The "live" portion of a trial occurs between the dashed lines — first patient, first visit (FPFV) on the left and last patient, last visit (LPLV) on the right. The cartoon depicts the relative time relationships of a whole host of other activities that, individually, are crucial to the success of an individual trial. Keep in mind that essentially everything represented in Figure 9-1 is repeated each and every time a trial is conducted in the development program.

9.1 Clinical Trial Monitoring

There can be several different people involved in the process of monitoring a clinical trial. Four individuals deserve special mention at this point.

A **medical monitor** is someone responsible for the medical conduct of the study. This is a physician (the person might be an employee of the sponsor, or a CRO contracted to conduct the study, or an independent individual) who reviews, on a regular basis, laboratory data and adverse event reports, and answers questions regarding application of inclusion and exclusion criteria among other activities. Ultimately, the medical monitor is responsible to the sponsor.

The person designated as the "monitor" for a study usually goes by a title of **Clinical Research Assistant** or **Clinical Research Associate**. Those titles are usually abbreviated as CRA. (This person also may be an employee of the sponsor or of the CRO responsible for conducting the study, or they may be an independent contractor.) Again, ultimately, this person is responsible to the sponsor. In general, the activities of a CRA fall into study planning, study preparation, study initiation, conduct and close-out, and study wrap-up.

At clinical trial sites, the **principal investigator** (who is frequently assisted by sub-investigators) is responsible for the conduct of the trial and care of the individual trial subjects (at least with respect to the trial). The investigator must be qualified "by training and experience" to conduct the study in the population selected for inclusion in the study. As a result, a podiatrist wouldn't be the principal investigator in a congestive heart failure study, nor a cardiologist the principal investigator in a foot ulcer study.

Also, at the clinical trial site, a **clinical research coordinator** (CRC) works under the direction of the principal investigator and handles the bulk of the daily clinical trial activities. The activities might range from evaluating new studies for feasibility, preparation for a new study to be conducted at the site, managing the conduct of all aspects of the trial, and assisting the sponsor or CRO with the close-out of the study. These folks are, in my opinion, of prime importance in the conduct of the study at the site and frequently make the difference between success and failure at the trial site.

9.1.1 Purpose of Monitoring

Clinical trial monitoring is performed in order to verify that the rights and well-being of human subjects are protected, the reported trial data are accurate, complete, and verifiable from source documents, and that the trial is being / was conducted in compliance with the currently approved protocol (including amendments), with the principles of Good Clinical Practices, and with the applicable regulatory requirements.

9.1.2 Responsibilities of Sponsor / CRO

It is the sponsor's job to ensure that trials are adequately monitored. Included in that responsibility is determination of the "appropriate" extent and nature of that monitoring. The determination is based on considerations such as the objective, purpose, design, complexity, blinding, size, and endpoints of the trial. As a general rule, monitoring is conducted at the clinical trial site before, during, and after the trial is conducted.

9.1.3 Requirements for Monitoring

The requirements for monitoring have been developed in several ways. For example, the US and the ICH have established a set of regulations, guidances and guidelines to define the monitoring requirements. In addition, the pharmaceutical and device industries have developed, over time, a set of standards that characterize the activities and procedures involved in monitoring.

9.1.3.1 United States (U.S.)

The US requirements for clinical trial monitoring are contained in the CFR (Code of Federal Regulations).

- 21 CFR Part 50
  www.accessdata.fda.gov/scripts/cdrh/cfdocs/cfcfr/CFRSearch.cfm?CFRPart=50

- 21 CFR Part 56
  www.accessdata.fda.gov/scripts/cdrh/cfdocs/cfcfr/CFRsearch.cfm?CFRPart=56
- 21 CFR Part 312
  www.accessdata.fda.gov/scripts/cdrh/cfdocs/cfcfr/CFRSearch.cfm?CFRPart=312

There are also FDA Guidances

- Guideline for the Monitoring of Clinical Investigations
  www.fda.gov/downloads/ICECI/EnforcementActions/BioresearchMonitoring/UCM133752.pdf
- Guidance for IRBs
  www.fda.gov/downloads/RegulatoryInformation/Guidances/UCM171256.pdf
- Information Sheets for IRBs
  www.fda.gov/ScienceResearch/SpecialTopics/RunningClinicalTrials/GuidancesInformationSheetsandNotices/ucm115834.htm

The study protocol also contains a section on monitoring that specifies the monitoring procedures for that trial. Generally, it is a boiler plate section — meaning, once it has been completed for the first protocol, it is usually unchanged (perhaps with minor modifications) through the rest of your program.

9.1.3.2 International Conference on Harmonisation (ICH)

The ICH Guidelines for clinical trial monitoring are contained in document E6, the GCP guidelines. (www.ich.org/LOB/media/MEDIA482.pdf)

9.1.3.3 Industry Standards

Industry standards aren't really written down in a single location. Large pharmaceutical companies, device companies, and CROs have established their own sets of procedures and will train their personnel on these procedures. When compared from company to company, the procedures are very close, frequently differing only in terminology and description.

In addition, other organizations, such as the Drug Information Association and Regulatory Affairs Professionals Society, conduct courses for beginning and more advanced CRAs. In those training sessions, detailed presentations of monitoring requirements are presented. Information presented in these courses obviously has its grounding in the United States regulations and requirements as well as the ICH Guidelines, but also incorporates what would be accepted as "industry standards".

9.1.3.4 General Responsibilities of Monitoring

Table 9-1 lists many (though not all) of the general responsibilities of a site monitor, the CRA, working on behalf of the sponsor of the trial. (Remember the monitor, the CRA, might be an employee of the sponsor, a contract person, or an employee of a CRO, but the responsibilities are the same.) A more detailed list is available in ICH Guidance E6.

9.2 Study Site Management

Study site management starts with site identification and moves through to the point of closing down the site.

9.2.1 Site Identification

How do you find investigators? There are a number of different ways to identify investigators and investigative sites.

Let's use hypertension as our target indication. In general, hypertension is treated by a large number and broad group of physicians. General practice physicians, family physicians, internal medicine physicians and cardiologists might treat it. Even pediatricians and gynecologists may treat hypertensive patients. So, we have a large pool of potential investigators from which to choose.

The first requirement for a potential investigator is that they be qualified by "training and experience". To use our prior example, the

Table 9-1. Monitoring Responsibilities

Monitoring Responsibilities
Serving as main line of communication between Sponsor and Investigator
Verifying that Investigator has adequate training and experience to conduct study
Verifying that Investigational Product has been adequately stored, distributed, and accounted for
Verifying that Investigator has followed the approved protocol (and amendments, if any)
Verifying that Informed Consent was obtained prior to each Subject's participation in the study
Ensuring that all current Essential Documents are at the site
Verifying that all Subjects enrolled in the trial meet inclusion / exclusion criteria
Verifying that Source Documents are up-to-date and accurate
Verifying that Investigator adequately satisfies all reporting requirements
Ensuring accuracy of all data recorded in Case Report Form (consistent with source documents)
Ensuring that all Adverse Events are accurately and adequately reported in accordance with appropriate regulations

podiatrist, however well trained he or she might be, is most likely not the ideal investigator for a hypertension study, and the cardiologist, again no matter how well trained he or she might be, is most likely not the ideal investigator for a foot ulcer study.

In addition, the potential investigator <u>must</u> have a facility and an organization (e.g., staff) within that facility that enables them to do clinical trials. Next, the investigator needs to have access to a large enough pool of patients fitting the requirements of the protocol. (Such patients might come from the investigator's practice or might be referred from other care givers.) In addition, prior experience with clinical trials is certainly helpful, although — in the right circumstances — not a requirement.

One place to look is the internet, believe it or not. Also, you can obtain some names speaking to other people who have done clinical trials, particularly in this therapeutic area. Sometimes you will identify an institution of interest in a geographical region of interest and then determine which physician or physicians at that institution might be potential investigators. Of course, if a CRO is involved, you might rely on their prior experience to identify effective sites.

Once you get away from a very specific common type of illness, like hypertension, the selection process for investigators becomes a bit more challenging.

As a different cardiovascular example, consider congestive heart failure. A broad range of physicians can treat congestive heart failure also, but, depending on your particular study, most likely you are going to require a specific subset of patients with congestive heart failure. The inclusion / exclusion criteria might limit the type of patient eligible to a very specific type of patient with CHF. As a result, your target set of investigators would be a group with experience treating that type of patient. Hence, a more sophisticated (in terms of disease management) investigator would be desirable. You probably want them to have a sub-specialty in cardiovascular disease. And whether that is an internist who has a sub-specialty in cardiovascular disease or whether it's a cardiologist is your own call, but my thought would be that, the more sophisticated the criteria for enrollment of patients, the more likely you are to want a sub-specialist.

Regarding investigative site locations, for diseases such as hypertension you definitely don't need to go to academic sites to try to recruit patients. Once you move into the realm of more specialized diseases, however — for example, we talked about congestive heart failure — you may want to identify academic investigators as potential investigators for a particular project or product.

For the more specialized study and patient, a good place to start is Medline (or PubMed; www.ncbi.nlm.nih.gov/pubmed/). Enter the disease and key characteristics and survey the last 2-3 years for publications on the topic. Identify all the articles you can find on your particular therapeutic area and then just simply look at the number of articles published by any individual. Look at their investigative sites and then go through a process of contacting them. That is one means of finding physicians who have experience doing clinical trials and, in addition, publish the results.

Another way is to look at sub-specialty physician directories. For example, if looking for a cardiologist, contact the American College of Cardiology. You can identify a list of fellows of the American College of Cardiology.

9.2.2 Site Qualification

Your first contact with the investigators is usually simply assessing interest. You might describe the study in very general terms, "We have a product we are beginning to evaluate in hypertension. We think it has a very interesting profile, because we believe it's safe, we believe it's effective when used once a day, etc., and our study would involve . . ." Then, you can describe the very basic sets of parameters for that particular study. The investigator ought to be able to tell you, "Yes, I am able to do this at this point in time." Or "Gee, I can't do it for the next two months because I am engaged doing something else" or "I've got some problems teaching or lecturing (or whatever it turns out to be)" or "Gosh, I just can't get to that. So, I don't think I'm interested."

If the potential investigator expresses interest, next probe the number of patients he or she sees in a (for example) month that fit your key inclusion / exclusion criteria. Always take the responses with a good deal of caution and skepticism.

(Whether the physician has said "yes" or "no" to your interest questions, ALWAYS ask for at least two recommendations for additional investigators. Trust me. You want to expand your list.)

Once one has assessed that level of interest, the investigators need to be specifically qualified. (See Table 9-2.) The initial stage of qualification can be done either in person or telephonically. In either case, it is important to use a formal check list (such as Table 9-2). The questions should be tailored to the specific requirement of the trial.

Table 9-2. Site Qualification Check List (prototype questions)

Investigator qualified by training, education, and experience?
Support personnel (sub-investigators, CRCs) are currently available and will be available through the duration of the study?
Site has adequate resources, staffing, and facilities?
Site has adequate (and appropriate) storage facilities for study supplies and clinical trial material?
Investigator understands protocol conduct and requirements?
Site can recruit (from its own practice or through referrals) patients who will meet the protocol's inclusion / exclusion criteria?
Local laboratory (if used) has current licensure and accreditation?
Research activity flow and patient handling are "reasonable"?
Investigator and all staff understand (and agree to) requirements of GCPs?
Institutional review procedures, including submission procedures and timing and review and approval cycles are known?

If the sponsor and the sponsor's team have not conducted a trial in this therapeutic area previously or there are some unique aspects to the trial, it might also be useful to identify a couple of experts in the field to determine some important characteristics of investigators and investigative sites that might enhance the chance of success in the trial. Of course, the more operational questions (e.g., GCPs, supply handling, etc.) will remain the same and require specific querying.

Let's consider some of these topics in a bit more detail.

Familiarity and real understanding of Good Clinical Practices by the personnel to be involved in the study at the site is really important. Part of this involves information, but part of it is process as well. Does the site have the right type of clinical study coordinator? Is there good coordination with laboratory facilities? It there good data flow through the institution, e.g., reports of procedures and diagnostic studies? Are there adequate facilities for storage of records and supplies?

The patient population to be included in the study must be discussed very specifically with the investigator and the site. The first question, using our target of hypertension, would be, "Do you see hypertensive patients?" If the answer is in the affirmative, then explore hypertension within the context of the specific inclusion / exclusion criteria. Make sure the investigator fully understands the target population. Once that is clear, explore the available pool of patients. Would they come from the investigator's practice? Would they come via referral from inside the institution? Would they come via referral from outside the institution? If referrals are going to be utilized, how will contacts be made? Will advertising be required?

Following that discussion, attempt to determine how many patients would really be recruitable for the study. In trial planning, it is ALWAYS best to be conservative in enrollment projections.

To turn the coin over for a moment, just because somebody says they have plenty of the patients of interest, it does not necessarily mean that this protocol will enroll. While hard to ascertain over the telephone, it is important to understand the subtle personal dynamics at the site. In looking at work flow, does the site appear to be well organized? Do the staff appear to be cooperative and responsive to each other? (Is there a sense they are cooperative and respectful or is there a suspicion they actually hate each other?)

One other key item: Is the site running a competitive study at the same time? For example, if the site is already participating in a hypertension study being conducted by another company, that study needs to be put into perspective. If the other company's study is finishing up and will not be running concurrently with this study, that is good. However, we need to make sure the remaining pool of patients is sufficient for our study. A second option, however, is that the study *will* be running concurrently. Now the decision point turns on the inclusion / exclusion criteria. Do they overlap? If so, perhaps our study only gets the "left over" patients — not good. Or, perhaps, the criteria don't overlap, but now our question has to do with resources at the site — can they run both studies effectively? These are very important issues to resolve at each site that might be used for the study. The challenge is to get honest information from the site.

I have certainly been in situations in which people have been less than honest about participation in competing studies. One approach to the issue is to write language regarding competing studies into the investigator agreement. "I am hereby certifying at this point in time that I do not have any competitive studies" or some wording such as that. (Run that by the lawyer, of course.) The goal is to receive assurance from the site that they will expend honest effort to enroll the study. Sometime the signature on the piece of paper helps.

In the querying process, also understand that the sponsor needs to know about institutional review procedures (IRB and any other committees), specifically understanding submission requirements, timing of submission, frequency of meetings, expected time for review and approval.

9.2.3 An "Ideal" Site

As a reference point, what would an ideal investigative site look like?

Of course, the site would feature an experienced investigator with adequate staff who are fully trained in good clinical practices. Their population pool meeting the trial's inclusion / exclusion criteria would enable them to fully enroll their site's quote in the first week the protocol is open. There is great work flow and organization. All patients who are enrolled in the study complete the study. All of them have perfect records.

While unlikely to happen in real life that way, the description could serve as a benchmark as potential sites are considered.

I noted that it is important to identify an experienced investigator. Does that mean no naïve site would ever be considered? The answer is, "No." But, if such sites are used, the levels of scrutiny, anxiety, and concern about compliance, monitoring and so forth will certainly be heightened. I would suggest in new organizations, i.e., groups that have never done clinical trials before as a group (individuals may have done them in another company, but the group hasn't done a study together), it might not be the best time to try brand new investigative sites.

9.2.4 Investigator Meetings

Once the investigators and sites have been identified, it is time to get them all together for an Investigator Meeting (IM).

9.2.4.1 Purpose

I believe it is important to have a face-to-face investigator meeting that involves the investigators AND the clinical study coordinators for each of the sites at the start of any multi-center study. Generally, it works best at a central location (as examples, if in the US, favorite locations include Chicago, St. Louis, and Dallas; for studies in Eastern Europe, Vienna and Prague are relatively convenient locations; for studies in Asia, Bangkok seems to work fairly well). The duration of the meeting is dictated by the content and the willingness of the investigators to attend. A day and a half to two days is generally a good time. In planning, factor in travel.

9.2.4.2 Advantages

One key advantage is that everybody who participates in the study gets the same story at the same time. They've heard the same questions and the same answers. It is ideal that each investigator and his or her study coordinator be at the meeting and go through the whole process as well. Key point: Try as much as possible to discourage the late arriver / early departer participant. It is important that everyone receive the same message — part of that is hearing it at the same time. That goes a long way to reassuring regulatory agencies that in fact everybody conducted the study the same way. Ultimately, poolability of data is an important consideration. One of the first steps to ensuring that result is to conduct investigators' meetings attended by ALL investigators and coordinators.

There are generally multiple goals at an IM. Some are operational and some are strategic. Operational aspects are relatively straightforward, and include items such as protocol review, handling of laboratory specimens, details of case report form handling, and handling diaries and randomization systems. Discussion of GCPs also falls into that category.

On the strategic side will fall discussions of the protocol — patient population, endpoints, visit structure, data collection, and baseline standard of care. There are generally two approaches to these discussions. First is the IM in which the protocol is presented to the investigators "AS IS". The second is the IM in which the protocol is "open for discussion". Each type of meeting has its purpose. However, do not mix them!!!

If the protocol is already set, the only modifications would be fixing typographical mistakes or minor modifications for clarity. This type of meeting is really educational. "This is how the protocol will be conducted." If an investigator can't live with the protocol (although ideally these objections would have been identified before the investigator ever got to the meeting), he or she would need to be excused and dropped from the protocol.

In the discussion-oriented IM, the goal is to generate investigator agreement on key operational points. This is a challenging format and requires the moderator to have experience leading and resolving brainstorming sessions. For this type of meeting to have its greatest value, resolution must be achieved for discussion points ON THE SPOT. At the end of the meeting, there is explicit agreement among the participants on key point and key wording. Again, it is possible an investigator can't live with the group's decisions. If true, that investigator, hopefully, departs on good terms, but they would not be included in the study.

The decision regarding the respective formats is dependent upon the product, the company, the protocol, the investigators, the timetable, etc.

While I generally believe that the most effective investigator meetings are conducted face-to-face, in today's world there are other options. My belief is that the initial investigator meeting ought to be in person, ought to be live, and ought to include all of the investigators and the study coordinators. There may be follow-up investigator meetings in which updates of information, opportunities for problem resolution and inter-investigator discussions are provided; these are particularly useful if specific issues have arisen during the course of the study. Those types of meetings could be handled using facilities such as Net Meeting or via teleconference, or videoconference of some type. Nevertheless, I believe that there is a lot of value to having a face-to-face initial investigator meeting.

9.2.4.3 Timing

The timing of the investigator meeting is somewhat dependent on all of the sponsor's objectives with the meeting. The timing is ideally in the vicinity of four to six weeks before the study gets kicked off. Any closer to study start and it is difficult to make adjustments, if there need to be adjustments, particularly protocol modifications. Too early and people don't remember what was discussed.

9.2.5 Study Initiation

Site monitoring includes three sets of visits to insure the trial is conducted properly. First is a site initiation visit, followed by routine monitoring visits, and, in turn, followed by site close-out visits.

Once the site is identified and the sponsor is satisfied with the quality of the site and the likelihood of producing an adequate number of patients, a site initiation visit needs to be conducted. During this visit, all of the operational aspects of the study are reviewed, including details of the protocol, handling of patients, trial supplies, etc. — in short, all of those items covered under sponsor and investigator responsibilities in the discussion of Good Clinical Practices (Chapter 4).

Occasionally, a sponsor may — with an experienced site — handle the initiation at the Investigator's Meeting. As noted above, the operational information is handled, in any case, in detail at the IM.

The level of professionalism that the CRA brings to this particular aspect of conduct of clinical trials will translate to the level of effectiveness in the conduct of the clinical studies. The approach to the site, whether very experienced or less experienced, is essentially the same. In either circumstance, if the initiation visit is conducted in a professional, high-quality manner, the study is more likely to be successful. If the site is experienced, the phrase, "As you know . . ." might be used frequently, but the necessary information is still reviewed. On the other hand, if the site is less experienced, the meeting will be conducted in the same professional manner, but may be more directed to education than to reinforcement.

9.2.5.1 Preparation for Visit

Arrangements for the visit should be made early enough to ensure attendance at the site of the key personnel, including the investigator. Once the date and time have been set, I would suggest sending a confirmation email (ensure there is a response from the site) or calling to confirm the appointment.

Develop an agenda that is shared with the site ahead of time. The agenda should be clear about the time required and the expectations about attendees for specific sessions on the agenda. For example, the investigator might only be present for 1–2 hours in the morning portion and then return for an hour in the afternoon part. Make sure there is prior agreement to each item on the agenda, the attendees and the timing.

Also, the sponsor should have its own check list and script to focus the conversations. The check list is important — it should conform to the SOPS (sponsor's or CRO's) in place — to ensure that all items are covered during the discussion. (As an example, see Figure 9-3.)

Prior to the visit, rehearse. Even if the CRA is experienced, I would suggest walking through the visit, what will be covered in each section, noting in particular the key points. Particularly, ensure that the key points will be covered in the time allotted.

9.2.5.2 Conduct of Visit

Table 9-3 lists some of the items to review during the visit.

During the initiation visit, the monitor will ensure that files have been set up to contain all of the Essential Documents. (See Chapter 5.)

It is generally better that the sponsor has its own format (or the CRO's format) and simply put that in place at the investigative site. As a general rule, I believe it is best to provide a set of pre-marked hanging files or 3-ring binders for use by the site.

One item not discussed yet as an essential document is the FDA Form 1572. In the US, the Form 1572 is required. Many other countries in the world use the Form 1572 for the same purpose, although it doesn't carry legal weight. There are some countries which prohibit use of the Form 1572 because of the legal language referring to US regulations on the form, although a similar form may be used for the same purpose. The intent of the form is to identify the investigators who will be conducting the study and to require them to acknowledge the following statements: "I will conduct this study to the best of my ability. I will do it honestly. I will do it according to the protocol." It is important to have a curriculum vitae (CV) for each person signing the Form 1572 (or similar form)

Studies may utilize the services of a local laboratory (at or near the institution), a central laboratory (a large commercial laboratory with the ability to transmit results electronically to sites and the sponsor / CRO), or, occasionally, both. In all cases, the laboratory's range of normal values needs to included in the site files. Also included will be laboratory certifications and accreditation certificates.

The IRB roster, along with duties and qualifications of each of the members of the IRB will need to be included in the site files. The protocol on file needs to be the current version with a completed signature page on the front of the protocol. Additional documents that are deemed important by the sponsor or the CRO will also be included in the file of essential documents.

Discuss in detail the handling of all study supplies such as case report forms, diaries, clinical supplies and storage, laboratory supplies, and any other materials related to the study. A particularly critical element in the discussion is the handling, processing, dispensing, and return of the test product (active and placebo) supplies (accountability).

Table 9-3. Site Initiation Visit Check List

Investigative Site Document File
Signed FDA Form 1572 (in US)
Investigator's CV
Signed protocol (+ amendments, if any)
Investigator's Brochure (current)
IRB information (membership list, protocol approvals, including consent forms, advertising approvals)
Personnel signature list
Laboratory certification (current) and reference ranges – if local lab used
Study and Protocol Review
Rationale for test article development
Background and purpose of study
Key available data, including pharmacology, toxicology, pharmacokinetics (Investigator's Brochure)
Inclusion / Exclusion criteria
Enrollment procedures
Study treatment procedures, by visit
Procedures for withdrawals
Special tests and procedures and ordering supplies
Good Clinical Practices and Regulatory Compliance
Investigator obligations
Informed consent procedures
IRB / IEC reporting requirements
Sponsor reporting requirements
Study record storage requirements
Adverse Event Reporting
Potential reactions and management
Reporting requirements and procedures
Serious adverse event reporting requirements and procedures

Table 9-3. Site Initiation Visit Check List (continued)

Case Report Forms
Instructions for completion
Typical problems or errors
Query process
Correction process
Test Article Control
Receipt
Storage requirements
Accountability
Dispensation procedures
Reordering of supplies
Return of unused supplies
Study Schedule and Targets
Enrollment targets
Monitoring schedule
CRF submission deadlines

Use this opportunity to review the protocol. Highlight the key items (don't go through every page and every item). Review in particular the inclusion / exclusion criteria, any specific "tricky" items in randomization processes, unique items (such as questionnaires or unusual processing of lab samples), and very specifically and clearly review procedures for safety reporting.

A lot more information is covered, but it is important to make sure these items, at least, are covered specifically. Next, document the visit.

9.2.5.3 Follow-Up of Visit

Finally, at the end of the meeting you will follow up and provide a thank you letter. Of course Dr. Famous and each of the staff who participated are acknowledged. It is also important, in this letter, to address any issues that need to be resolved subsequent to the meeting. The issues should be specifically identified, the expected resolution, and the time by which the resolution is required.

If, at the end of the site initiation visit, there are any signals suggesting this is a potentially problematic site or if necessary elements are not in place, STOP! Do not proceed unless and until the issues are resolved. Communication at this point needs to be clear and very specific. It should be in writing (in a letter) and delivered via certified mail or another means requiring a signature upon receipt.

9.2.6 Site Relations / Routine Monitoring Visits

The next set of visits is the interim or routine monitoring visits for the study. The frequency and number of those visits are directly related to the length and complexity of the protocol. In general, one likes to visit a site after the first one or two patients are enrolled to ensure that procedures are being followed and CRFs are being completed correctly. Additional visits could be every third week, or every month, or even less frequently. The visit schedule will depend on the duration of the study as well as on the rate at which patients are enrolled.

Note (if you have never conducted a clinical trial before): In general, the initial routine visits tend to be more frequent. Specific items of interest are adverse events, CRF completion, validation of CRFs compared to the source records, etc. With longer protocols, visits become less frequent during the middle of the study. For example, a hypertension study would typically go on for three months, potentially as long as six months. Congestive heart failure also might go on for a longer period of time. With longer studies, the intervals between study visits tend to be longer (e.g., monthly or bi-monthly, instead of weekly). This type of schedule would likely translate to less frequent monitoring visits as well.

The specifics of study visits are not identified in regulations. Thus, the spnsor's (or CRO's) SOPs determine the frquency and activitis of thes vist*Sidebar: Monitoring and analysis in larger studies.*

> Sometimes, if there are concerns about the mechanics of the protocol in actual operation, such as procedures or sequence of activities, it is worthwhile closely examining data from the first few patients. This might simply mean entering the data in the database very early to test reports with real data without breaking the blind. On occasion (rarely) the blind might be broken for these patients – realizing they will be handled differently in the final analysis – as they are entered in the database. The point of the exercise is to make sure the operation of the protocol will facilitate a successful outcome to the study. If the protocol needs revision, this is the time to do it. This process, if employed, is clearly written into the protocol at the beginning.

During that monitoring visit, review the case report form with the site personnel and discuss how the conduct of the protocol has gone. Was the protocol executable as written? Should there be modifications to the protocol to add, delete, or change procedures or the sequence of activities that will significantly improve the study's conduct and the value of the data obtained? If there turn out to be problems with the protocol, it is best to find them out early in the study so changes can be made without negatively impacting the data from a large number of patients.

9.2.6.1 Visit Activities

Look at Table 9-4 for an example of a check list for the monitoring visit.

During the visit, the CRA reviews all the case report forms and verifies the entries against the supportive source data. The personnel at the site enter data onto case report forms (CRFs). The actual CRF might be in a paper format or electronic. Data entered onto the CRFs derive from source documents. Source document information might be the physician records, laboratory data from the hospital or a central laboratory, electrocardiographic reports, x-ray reports, etc. During the monitoring visit, it is important that the CRA be able to identify corroborating documentation for every item entered in the CRF.

In addition to data verification, protocol compliance should be assured. The CRA must be thoroughly acquainted with all aspects of the protocol — activities, procedures, follow-up, study drug handling — and then make sure that the protocol is being followed correctly. This includes verification that each patient enrolled in the study meets ALL inclusion / exclusion criteria and that there is a signed, correctly-dated consent form for each patient undergoing any study-specific procedures. The most effective means of doing this is to proceed methodically from Patient 1 to Patient 2 to Patient 3, etc.

If the trial has already been going on for a period of time and data have already been provided to the data group, queries regarding the data may have been generated. Examples of queries might be: "The entry on page 12 is illegible"; "The information on page 12 and page 6 appear to be in conflict"; "Confirm the hemoglobin value on page 19 (it is out of range)." Resolution of queries not previously addressed is also an important activity of the monitoring visit.

The CRA, during the visit, will also review study enrollment (see Section 9.2.6.4, below) and drug accountability. The latter activity involves making sure storage conditions are appropriate, the appropriate study drug is dispensed to the appropriate patient (according to the randomization schedule), and the quantities of study drug supplied to the site equal the supplies dispensed plus the supplies on hand.

Consult Table 9-4 for other topics to cover during the visit.

At the conclusion of the visit, review findings (good and bad) with the site personnel. It is important that the investigator be present for that discussion. At the very least, these are good questions to ask: "How is our study going? Do you see any problems? Do you see any issues that we ought to be thinking about as the sponsor, as we look to conducting the study in multiple investigative sites in other locations?" Occasionally, reality intervenes, particularly for investigators with active clinical practices, and the investigator won't make the review discussion. In that case, a telephone conversation might have to do. (If the investigator repeatedly misses these meetings and / or the staff appear uncooperative, this should trigger alarms for the CRA and the sponsor / CRO.)

Table 9-4. Routine Monitoring Visit Check List

Investigative Site Document File
Signed FDA Form 1572 (in US), all current investigators?
Investigator's CV, current?
Signed protocol (plus amendments, if any), most recent version?
Investigator's Brochure (current)?
IRB information (membership list, protocol approvals including consent forms, advertising approvals), any additional correspondence or changes?
Personnel signature list, new personnel?
Study and Protocol Review
Changes in staff?
Changes in facilities?
Protocol requirements being followed?
Inclusion / exclusion criteria met for all enrolled patients?
Consents for all enrolled patients (dated before entry)?
CRFs reviewed? (Issues discussed with staff?)
Source documents reviewed?
Queries resolved?
Corrections made?
Protocol deviations?
Any communications with IRB?
Completed CRFs collected?
Recruitment on schedule?
Adverse Event Reporting
Any serious AEs since last visit?
If yes, were required forms completed satisfactorily and submitted?
Any outstanding events or forms?
Was IRB informed, if required?
Drug Supplies
Investigational product stored properly?
Dispensing procedures followed?
Investigational product accounted for?
Are supplies adequate?
Administration
Issues discussed with site personnel and investigator?
Appointment for next visit made?
Written report submitted?

Following the visit, the study report will summarize all of the findings — good and less-than-good. If there are action items to address, the visit report will identify the issue, the needed resolution and the time frame in which it is to be accomplished.

9.2.6.2 Site Communication

Effective communication throughout the study is crucial. This involves communication before the visit, during the visit and as a follow-up to the visit. Between visits, there also should be routine communication. Sometime this would take the form of periodic investigator telephone calls. It also might simply be CRA to coordinator communication. Another tactic to facilitate communication (and occasionally to induce healthy competition) is to distribute a trail newsletter that might inform the sites of changes in the protocol, new data, and, oh by the way, provide a listing by site of the numbers of patients screened, enrolled, and completed.

9.2.6.3 Documentation

Remember to document everything! Have a system in place to deal with phone reports / phone conversations. Determine where and how email correspondence will be filed. Develop an archiving system for monitoring reports and all correspondence. Also, importantly, ensure there is a system in place for recording all study exceptions / exemptions.

9.2.6.4 Enrollment Issues

Enrollment issues are traditionally the number one issue in trial management and represent, in general, the single biggest source of delay in the time line. For enrollment issues I would suggest using a problem-solving approach and try to work with the investigative sites. Work with them prior to the time the study starts, and frequently during the course of the study to try to find the optimal ways to recruit patients. (Preemptively you can consider hiring someone to work on advertising to encourage enrollment. Perhaps your CRO provides that service.)

The ideal study start includes having one or more sites that bring patients into the study right at the beginning. They have pre-screened their patient lists and have identified several individuals to contact as soon as the study is a "go" for screening. Hopefully, this leads to a burst of activity at the beginning. At the same time, the sponsor and the sites need to plan for enrollment beyond that initial burst. It is best to have developed that strategy before study start.

One set of options includes advertisements, either via radio, newspaper, television, etc. (Remember that content of those advertisements needs to be pre-approved by IRBs or IECs. See Chapter 7.) Another set of options includes "lunch-and-learns" for hospital staff to improve internal referrals. Additionally, communications — verbal or in writing — to externally-based referring physicians is helpful.

Understand FULLY the target population and how many evaluable subjects are being targeted. "Evaluable" is a word that should be explained. I am not sure it has made its way in to Webster's Dictionary yet, but it is commonly used in the industry. An **evaluable patient** refers to someone who participates in a study, adhered to the protocol and completed all study visits, and the data recorded for the patient are accurate, particularly regarding the key efficacy and safety endpoints. Evaluable patients are very valuable!!

As a study is planned, consider the following flow: Patients will be screened, but a certain number won't meet the inclusion / exclusion criteria for various reasons. Those who don't are now labeled screen failures. The percentage of subjects who fail in screening constitutes the **screen failure rate**.

Once a patient enrolls in the study, not all will complete — for various reasons. Those who don't complete are now labeled drop-outs. The percentage of subjects dropping out during the study constitutes the **drop-out rate**.

The math works in the following way: Start with the evaluable number needed and back-calculate the number who must enroll (using the drop-out rate) to product that many evaluable patients. Then, using that enrollment number, use the screen failure rate to determine the number who must be screened to, ultimately, produce the needed number of evaluable patients. For planning purposes, I would always add a bit of a fudge factor to that screening number, just in case . . .

As an example, assume 100 evaluable patients are needed for the study. Of those enrolling, approximately 20% are expected to drop out (for whatever reason) during the trial. In addition, approximately 25% of those screened will not meet the study's inclusion / exclusion criteria.

100 evaluable patients
  20% drop-out rate
125 patients needed to enroll in the trial
  25% screen failure rate
160 patients must be screened to produce 100 evaluable patients.

I indicated I would use a fudge factor at the end. As a result of these calculations, I might plan to screen 170-175 patients to ensure we would achieve our evaluable patient goal. It is important that the screen failure rates and drop-out rates be followed closely during the trial to make adjustments as needed. Specific factors that influence the overall recruiting include the study's complexity, the study's length, and the number of procedures.

If the study targets patients with an unusual or rare disease, it is important to work with the respective patient advocacy group. For example, a company doing a study recently in the lupus erythematosus area worked very closely with patient groups. Patients who have lupus are very interested in potentially new types of treatment for their particular disease. So, as a result, working with them might facilitate enrollment into the clinical trials as well.

Even with experienced sites and with pre-formulated strategies in place, enrollment is always an issue. It is important that the CRA / sponsor / CRO stay on top of it throughout the study. As a last resort, the protocol's inclusion / exclusion criteria might need to be changed to save the study. I say as a "last resort" because of the implications for data analysis and interpretation, but keep the option on the table.

Table 9-5. Sources of Study Conduct Delays

Percent of Delays (IND to NDA Approval)	
Total Delays	
Data analysis and reporting	20%
Study conduct	40%
Approval process	40%
Study Conduct Delays	
Data collection and coordination	28%
Study logistics and initiation	22%
Patient enrollment	50%

In addition to the obvious factors of inclusion / exclusion criteria and the presence of competing studies, other factors may influence enrollments. Some additional examples follow.

Geography is important, particularly if multiple countries are involved. The timing of regulatory and ethical committee / IRB approvals also impacts start times and then, potentially, enrollment. Your supplies can get caught in customs. Perhaps, at that time, your patient falls out of the screening window defined in the protocol and might have to start over. In that case, the patient might decline further participation.

Seasons are also important. If the start of your study is delayed and you start screening for psoriasis patients in the summer, it will be slow. They spontaneously improve (at least as a population) in summer.

Finally, *local issues* — social / political climates, strikes (have you ever done a trial in France or India? — frustratingly entertaining), etc.

9.2.7 Study Close-Out

In addition, a close-out or a termination visit is essential to tie up all the loose ends at the conclusion of the study. The Study Close-Out Visit is quite important. Logically, the CRA should prepare for this visit ahead of time. Make sure the person performing the close-out visit has a road map detailing all the items to cover during the termination visit.

This is also a good time to remind the investigator and clinical study coordinator, on behalf of their site, about their responsibilities regarding record and document retention. As with all other visits, appropriate documentation and follow-up are required.

Table 9-6. Close-Out Visit Check list

Investigative Site Document File
Study document file complete, per SOPs?
Study and Protocol Review
Final study report provided to the IRB?
All CRFs are complete and have been submitted?
All patient diaries are complete and have been submitted?
All Adverse Event follow-up is complete and documented?
All source documentation is in order?
All informed consents are filed?
Drug disposition forms are completed and submitted?
Study drug (if any) has been returned, per Sponsor's instructions?
All other study materials have been returned, per Sponsor's instructions?
Archive files have been assembled per procedures?

A note on sites or studies terminated early. Of course, most frequently, studies are terminated simply because the number of patients from the site met our goal or the study itself met its goals.

Alternatively, a site may be terminated because of poor performance. For example, the study is nearly 50% enrolled. Essentially every other site has 10-15 patients enrolled; this site has only one. Since an open site costs time, energy and money, it makes sense to at least consider early termination. Or, all sites might be terminated because there was a problem with the protocol itself or additional information came to light requiring discontinuation of development. In either case, the site(s) go through termination visits and follow-up documentation. For the poorly-performing site, close it and say, "Good bye." If sites were doing well, but the program needed to be canceled, do your best to discontinue on good terms, so the sites might be utilized for future programs.

At the end of the trial, always leave the door open for future business relationships, particularly with the site that did everything they said they were going to do. Years ago, in doing a study for congestive heart failure at SmithKline, we used a particular investigative site. We needed about 15 patients from this particular site; this site enrolled their 15 patients within two weeks.

I was pretty new at this clinical development thing, so I was excited. The monitoring CRA asked about enrolling an additional 15 patients at that site. My response, "Of course!!" Supplies took about a week to arrive and, within two more weeks, the site had enrolled the next 15 patients. As one might expect, the CRA asked about adding more.

My supervisor, for some odd reason, didn't seem quite so enamored of this situation. He mentioned — loudly — something about center effects and — more loudly — something about fraud. We quickly arranged to have the SmithKline compliance group visit the site. This turned out to be one of those times when the stars aligned and I (we) got lucky. All 30 patients were accounted for and the data were real.

We did have to cut the enrollment because of the potential center effect, but we maintained good relationships for future studies. I, however, did learn several important lessons about trails and trial monitoring in the process.

Auditing is a normal procedure for any study. Its importance lies in verification of patients, data, record keeping, subject protection, and study drug accountability. As a result, sites need to be aware (at the beginning of the study) that audits during or after the study are a possibility. When explained and discussed early, the audit becomes less of a punitive exercise (frequently the feeling of investigators and site personnel) and more of a good business practice. Of course, as noted above, there may be for-cause audits, but they are also a routine component of clinical trial conduct and management.

9.2.8 Post-Mortem: Lessons Learned

At the end of the trial, I would suggest the sponsor conduct a formal exercise to understand what worked and what didn't in the trial. Good project management comes from experience. It also comes from learning from that experience.

New therapeutic areas and larger, more complicated trials will always bring new challenges. To be efficient in project management, it is ideal to repeat as much as possible with each new study. To be effective in project management, improvement is always desired and re-invention is often required. Getting the mix right is what makes this so much fun!!

9.3 Budgets

First, we will briefly provide an overview of how money is spent in clinical research and then identify the budgeting process for a single trial (including clinical trial material, the investigator, the investigative site, patients / subjects, sponsor, protocol, and financial disclosure) and finally touch on contracts (including issues and metrics) with the clinical trial sites.

Table 9-7. Spending on Clinical Development

Phase 1	19%
Phase 2	26%
Phase 3	42%
Phase 4	13%

Source: PhRMA

Table 9-7 shows the distribution of clinical development expenditures, both domestically and abroad, for PhRMA member companies from 1980 through 2001 broken out by phase of development. Approximately one fifth occurs in Phase 1, about a quarter in Phase 2, and nearly half in Phase 3.

Understanding this distribution is important as one considers the budgeting process for the development program and understanding the funding requirements for various stages. The distribution explains why it is common to look for partners before entering Phase 3.

Table 9-8. Where Does the Money Go?

Dedicated research centers	19%
National SMOs	7%
Academic medical centers	23%
Others	51%

Source: Center Watch

To whom does all that money go? (See Table 9-8) About 20% of the budget goes to dedicated research centers (places where research is their whole business, e.g., Phase 1 units). There are also site management organizations (SMOs) throughout the United States. Some of the money goes there. About a quarter of the money goes to academic medical centers. Twenty years ago, the share distributed to academic medical centers would have been closer to 50%. As clinical practices have changed with a greater focus on clinical research even in office-based practices, more of the clinical trial activity has shifted to those locations.

Figure 9-2 identifies the key components to be considered in the budgeting process.

9.3.1 Clinical Trial Material

In order to conduct a trial, clinical trial material is needed. Clinical trial material generally refers to the study drug and placebo. (It can also include other clinical trial supplies, such as case report forms, laboratory supplies, diaries, etc.) A whole series of activities go into the production and delivery of material to the site and, then, in turn to the patients. At least the following items and processes need to be accounted for: raw material, processing, analytical methods, packaging, coding, labeling, distribution, collection, and destruction (among others).

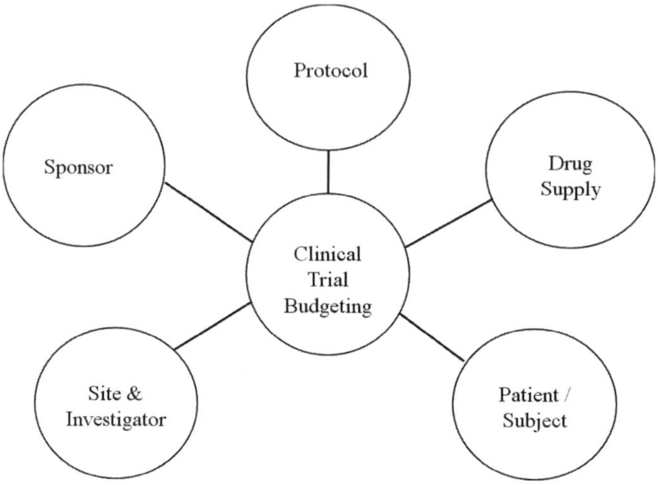

Figure 9-2. Major Components to be Considered in Budgeting

In smaller companies, this area might be one of the most challenging to work through and to get the right product to the right place. If expertise is not available in the company, don't try to wing this area — find help.

9.3.2 Investigator and Investigative Site

There are multiple costs associated with investigative sites as well. Institutional overhead is a significant cost. All academic centers impose very significant taxes (I don't know what else to call them) on all clinical research conducted at the center. The percentages range from perhaps 40% to 100% (really!). There are also IRB and Ethical Committee fees. There will be a fee for the investigator (for his or her time) and most sites will include other personnel (e.g., study coordinator) in the budget.

In general, budgets of site payments will need to include payments for screened patients, enrolled patients and completed patients. They will be allocated different amounts. Obviously, payments for completed patients are higher. However, investigators need to be incentivized to screen patients for entry and it is appropriate for the sponsor to pay for the site's time and effort. Enrolled patients who drop out during the study also will have a price tag associated with them. Determining the payment for the screen failure and the drop-out, as a percentage of the completed patient budget, is a bit more challenging to assign.

9.3.3 Patients / Subjects

Costs associated with getting patients into a trial begin with recruitment fees. These could be print ads or radio or TV spots. Other mechanisms are lunch-and-learns. In this format, the investigator / sponsor provides pizza, etc. and then reviews the disease and the protocol to educate other physicians, house staff, etc. about the protocol. Occasionally, physician referral fees may be used. The latter two techniques have been a bit more of a challenge with concerns about pharmaceutical company influence, but, if used correctly, can be helpful.

Transportation costs associated with the patient's travel to the medical center or office might be reimbursed so that patients are "cost neutral" regarding participation in a trial. Particularly for longer, more complicated studies requiring more patient visits, reimbursement for travel costs improves the chances that patients will be able to complete the required protocol-associated procedures and visits.

9.3.4 Sponsors

Occasionally, companies don't adequately consider their own costs. It is easy to understand checks that are written to outside vendors, e.g., CROs, consultants, regulatory agencies, ethical committees, etc. However, it is also important to consider the company's operational costs as part of the cash requirements of a trial or of a development program. For example, internal costs that might not be allocated to the project (and thus not fully accounted for) include clinical personnel, pharmacology studies and personnel, toxicology studies and personnel, pharmacokinetic studies and personnel, regulatory costs (personnel, if internal, and filing fees), all aspects of chemistry, manufacturing, and controls, and administrative overhead (all the other parts of the company that are supporting these clinical trial efforts).

9.3.5 Protocol

Specified in the text of the Protocol are the types of patients to be screened and enrolled, the visits they will have, procedures to be conducted, and assessments to be performed. Each of those activities costs money, whether as a specific cost — the cost of an ECG, the cost of the laboratory test — or as a time cost — the cost of administering the questionnaire, the cost of measuring the patient's vital signs. As you consider adding items because of "curiosity," note how much each will cost before blithely including it.

At the beginning of the planning and budgeting process, determine the following items, since they will have direct bearing on the overall costs: the patient population (how accessible, will advertising be required, will recruitment be slow — extending the company's operational costs), the number of patients required at the end of the study (for statistical purposes), the screen failure rate (how many people will undergo initiate testing, but fail to qualify for the study), drop-out rate (how many people will start the study, but fail to finish for whatever reason — longer studies and sicker patients are generally associated with higher drop-out rates), the number of centers required to accumulate the patients, the primary and secondary endpoints (and testing associated with determining endpoints), the number and timing of visits (obviously, the more visits, the higher the cost), and the number, type, timing and frequency of procedures to be performed for each patient during the study.

Once answers are available for these questions, a procedures and timing grid, as in Table 9-9, can be created. This type of grid should be included in the protocol as a handy reference for study personnel, and it serves as the basis of the budgeting process. It should identify ALL visits and ALL procedures, activities, assessments, or measurements that are pertinent to your protocol.

Using the protocol study grid as a template, each item — visit, test, procedure, etc. — is listed. To calculate the costs, determine the number of times the item is checked on the grid and multiply it by the unit cost for the item. Add each component together and the first set of protocol costs is available.

It might be surprising — even shocking — how quickly the costs add up. Table 9-9 lists some of the items (not nearly exhaustive) that might be included in a clinical study. In addition to the costs associated with conduct of the Protocol, it is also important to determine the study start-up costs, e.g., IRB / EC submissions, regulatory filings and maintenance, travel to sites, etc.

Once the total costs have been tabulated, time needs to be factored in. This is an area about which the company's financial person will care profoundly — cash flow. (See Table 9-10.) When will all these payments be made? What will trigger payments? Will payments simply be sent on a regular basis (e.g., once a month, once a quarter, etc.) or are they triggered by milestones (e.g., the 10th patient at a site, the 5th completed patient at the site, etc.)?

Table 9-9. Protocol Visits and Procedures Drive Costs
Protocol Study Grid

Procedure	Screening	Visit 1	Visit 2	Visit 3	Visit 4	Final Visit
Informed Consent	X					
Physical Exam	X				X	X
EKG		X				X
Vital Signs	X	X	X	X	X	X
Hematology		X		X		X
Chemistry		X		X		X
Quality of Life Questionnaires		X	X	X	X	X
Bone Density		X				X
Activity of Daily Living		X	X	X	X	X
Patient Diary		X	X	X	X	X
MRI	X	X				
Drug Dispensing		X	X	X	X	X
Adverse Events			X	X	X	X

As a general suggestion, consider setting up as many payments on a milestone basis as possible, in order to pay for performance. Another aspect not considered often enough are penalties for poor performance and rewards for good performance. One needs to be careful about these payments so that they don't have unintended effects — demotivators or motivating for the wrong reasons. In my opinion, both penalties AND rewards are important. If the site (or your CRO) exceeds your expectations — better enrollment, faster enrollment, cleaner data, etc. — give them a reward. It is best if they operate as partners rather than simply contractors.

Table 9-10. Payment Schedule (Cash Flow)

Start-up
• Initial site (or CRO) payments
On-going
• Milestone payments
• Additional payments
• Drop-outs / Follow-up visits
• Unscheduled visits
Study Completion
• Follow-up on AEs
• Study close-out
• Database lock

Tools to assist in the process of budget management are very important. Most CFOs have done this before in some other environment, so will have tools available to assist in the process.

However, consider using some specific tools, such as pricing the components of study budgets (procedures, tests, visits, etc.). Before going to the individual sites or to the CRO for their estimates, it is helpful to get some "ballpark" estimates of these figures.

One useful site is the VinCRO site (http://www.vincro.com/costestimator.htm). I have no vested interest in VinCRO, but they provide the service, so check it out. In addition, check with friends and colleagues and get estimates from CROs if you are using them. At the end of the process, the live costs from the sites, the CROs, etc. are, of course, needed, but this is a useful place to start.

9.4 Contracts

While there are multiple topics to consider regarding contracts and this book isn't really long enough to go into enough detail here, we will focus on some of the important issues to keep in mind in developing contracts for various vendors.

The respective contracting parties need to be clearly identified. Generally, this is straightforward, but occasionally there might be surprises; just make sure it is clear. Confidentiality terms need to be included. Most of the protection, from the sponsor's point of view, involves sponsor information, but the confidentiality aspect of the agreement is generally bilateral.

Termination clauses need to be clarified and specified. Mutual indemnification, except in the case of malfeasance, is a standard set of terms. As mentioned above, consider including appropriate rewards and penalties in the contract. (Such terms are not appropriate for all contracts, of course.)

Who owns the intellectual property (always the sponsor's, except when it isn't) needs to be clearly spelled out. Who, and under what circumstances, has the right to review and to publish also needs to be clear. I agree with the trend that all research should be published somehow — even the negative results (they put the positive results in context), but you, as the owner of the compound, want to have some control over the process and the content, most particularly in protecting intellectual property.

9.4.1 Principles for Site Contracts

In setting up contracts, consider some of these basic points. Ideally, payments should be based on performance. An up front payment makes sense; there are start-up costs with each vendor; but sequent payments should be tied to meaningful milestones.

As discussed above, payments for screen failures and drop-out must be included in the budgeting process. Nevertheless, since abuse can creep into these situations, the sponsor needs to monitor them carefully. It is also important to appropriately compensate the site for completed and "evaluable" patients; after all, this is what you really need to have a successful study.

In general, standardize as much contract language as possible. Standardization cuts down legal expenses and makes the process easier to manage.

9.4.2 Agreement Letter

This is really another contract. It includes a schedule of payments and should, in my opinion, clearly outline performance metrics (expected performance clearly defined; bonus or penalty for great or poor performance). When considering bonuses or penalties at investigative sites, one needs to be really clear that the response (bonus / penalty) is appropriately tied to the important metric for this particular study. This can be a particularly sensitive issue for individual sites (in contrast to CRO performance where setting bonuses / penalties is more common) and, if put in place, must be very clear to avoid conflict about achievement or non-achievement of the milestone. It is also important that the payment amounts and schedule be "reasonable" considering the work done.

9.4.3 Financial Disclosures

Beginning in 1999, all Phase 3 studies conducted in the United States needed to have documentation of financial disclosure. Subsequently, in the US and around the world, all trials require financial disclosure.

In the United States, the requirements are set forth in 21CFR 54. Disclosures regarding stock or equity interest in the sponsor company and cash payments directly to the investigator above a threshold are required. The whole intent is to avoid scientific bias and putting patients at risk in the context of biased research.

Who needs to provide financial disclosure? Any person listed as a principal investigator or sub-investigator on disclosure forms (e.g., FDA Form 1572) must provide financial disclosure. By the way, disclosure applies to family members as well. For example, the investigator might not own stock directly, but family members or family trusts might; that information must be disclosed. The reason why? It's fairly obvious. It's that the documentation needs to be in place such that an assessment might be made as to whether the compensation for the investigator or his or her financial interest could potentially influence the outcome of this particular study.

9.4.4 Clinical Metrics

Some useful metrics for clinical trials include: enrollment rate and number, screen failure ratio, drop-out rate, compliance rate, data clarification forms (data queries), protocol adherence, data quality, and documentation quality. Metrics related to contracts themselves are also important to include. Examples include execution time, renegotiation cycles, competitive budgets (other sites / other studies), regulatory or IRB / EC packets and processing times. What isn't measured can't be improved!

Whatever metrics are used, make sure the metrics used and the standards set make sense for the patient population and the protocol, and are meaningful in terms of important milestones. As metrics are set for the study (or for any activity, for that matter), determine the organization's expectations — transform the important objectives into milestones and time lines. Then, constantly assess progress against those expectations. As progress is monitored, problems can be identified early and rectified.

As a reminder, use budgets (and expenditures against budget) as indicators of performance. One important consideration is to benchmark this study against other studies the organization has run, as well as against those conducted by other organizations. Center Watch is a good source of data for benchmarking (www.centerwatch.com). The process should always allow for (and actually encourage) corrective actions that result in support and improvement for the sites. By way of motivation, it certainly is reasonable to set stretch goals. On the other hand, it would be useful to benchmark the metrics (best — average — minimal) with other studies / companies before putting them in place. In any case, manage the study to the metrics designated.

As an example of useful metrics, consider Table 9-11. Looking at these numbers, it is a wonder academic centers stay in business and that anyone uses them for trials. However, there are other reasons . . . remember, the Dr. Famouses of the world can be quite important in the process of validating your product to the rest of the medical community. But, such metrics are at least useful for planning and monitoring progress.

Table 9-11. Average Days to Study Start-Up

Academic	
IRB	88 days
Budget	42 days
Contract	70 days
Average Total	105 days
Private Hospitals	
IRB	43 days
Budget	21 days
Contract	21 days
Average Total	44 days
Private Practice	
IRB	16 days
Budget	12 days
Contract	14 days
Average Total	20 days

Chapter 10

Connecting Data — From Marketing Application to Patient

Essentially, the key to any study, to any program and to any marketing submission is the data. Our discussion of data and data management in this chapter will start by identifying what is ultimately needed for the marketing application and then, working backward, looking at data flow from a single patient to a final clinical study report.

Figure 10-1 illustrates the overall flow of information in a development project starting from the Development Plan and ending with the integrated analyses contained in the Marketing Application.

Connections

Integrated Analyses derive from . . .

Systematic Analyses of Pooled Data from . . .

Databases derived from . . .

Case Report Forms that record data from . . .

Individual Patient data that derive from . . .

Design of Individual Clinical Studies that incorporate . . .

Statistical Planning deriving from . . .

The overall Development Plan

Figure 10-1. Data Connections and Flow

10.1 Marketing Application

The point of this whole exercise — product development — is that we desperately want (and need) for our product to get to the market. In order to do that, we must construct and submit a market approval application for review by the regulatory agency that can grant our desires. My belief, as usual, is to identify the end target and work backward.

Figure 10-2, in a cartoon format, demonstrates the flow of information (in more detail than Figure 10-1) from the earliest point of our process — our decision about a target disease and our plan to improve health by finding a better way to diagnose, cure, treat, or prevent the target disease — through the process of identifying targets and designating leads, and, finally, into the processes of non-clinical and clinical development to the construction of the marketing application. From the point we say we are " in development", everything we do is geared to filling out that marketing application.

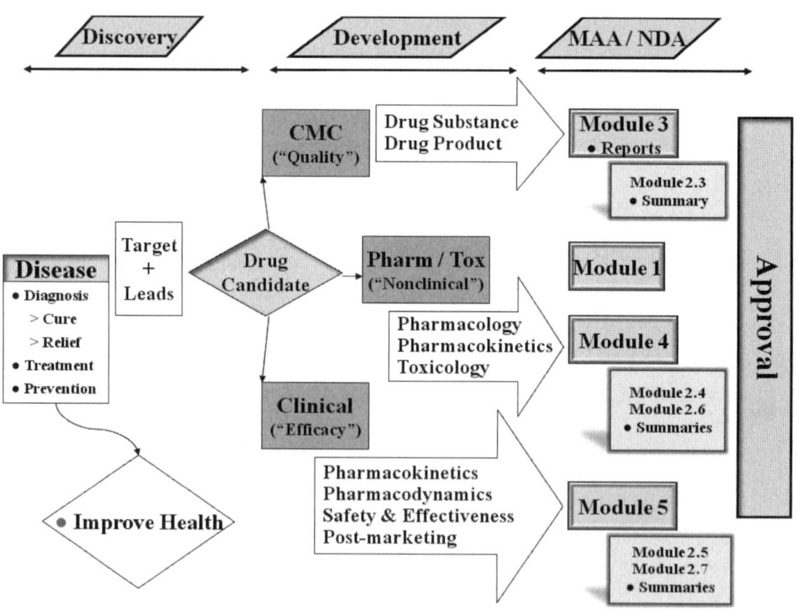

Figure 10-2. Development Information Flow

Currently, the Common Technical Document (CTD) is the standard for submissions in essentially all regulatory jurisdictions around the world — officially in some (US, Europe, Japan) and unofficially in many others. Therefore, this discussion will use the ICH format as an organizing principle. There are five modules in a CTD submission, cleverly numbered 1 through 5. See Figure 10-3. By the way, the overall organization of the CTD is described in ICH guidance M4 which is in its third revision (http://www.ich.org/LOB/media/MEDIA554.pdf).

Structurally, one can think of the Common Technical Document (CTD) as a pyramid with Modules 3, 4, and 5 forming the base — the structural support.

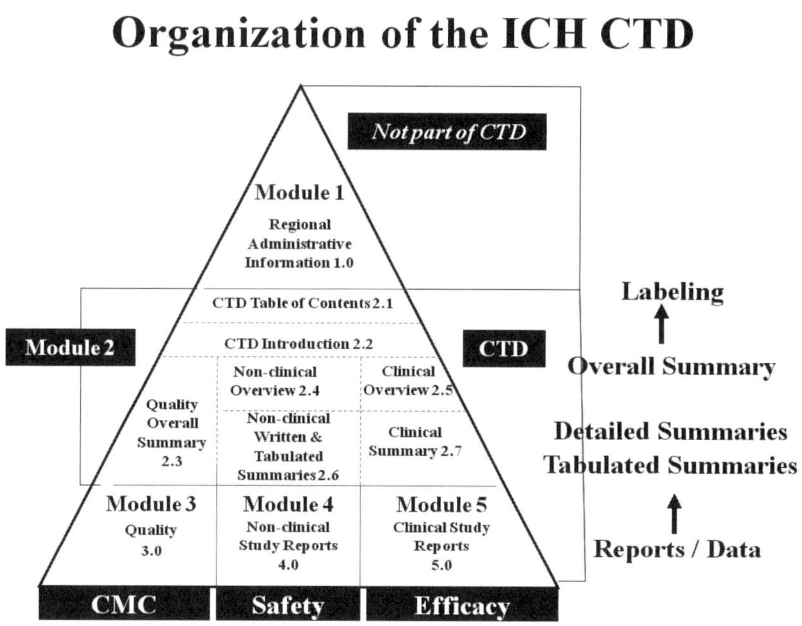

Figure 10-3. Organization of Common Technical Document

The components of Module 2 form the middle section and represent summaries of what is below. I tend to think of Module 2 as the "Sales" document. This is the document that is most mindful of the integrated assessment of risks and benefits.

Module 1, the capstone, is region-specific, the contents and structure of which are defined by the region or country to which the document is submitted. By virtue of its region-specific nature, it is not considered to be part of the CTD.

As noted, Modules 3, 4 and 5 form the foundation of the document. Module 3 is named Quality and contains the information about the drug substance and drug product, and all the information regarding manufacturing. (ICH Guidance M4Q (rl) is located at: http://www.ich.org/cache/compo/1325-272-1.html#Quality). Module 3 rolls up, in summary fashion, into Module 2.3.

Module 4 is called Safety but contains all the non-clinical studies we generally think of as being pharmacology and toxicology studies. (ICH Guidance M4S (r2) is located at: http://www.ich.org/cache/compo/1325-272-1.html#Safety). Module 4 rolls up into two summaries. First is a tabulated summary of the data below and is named Module 2.6. The overview summary, describing the importance and significance of the data below, is contained in Module 2.4.

Next is Module 5, the Efficacy section, that contains the clinical data and includes both efficacy and safety information from clinical trials. (ICH Guidance M4E (rl) is located at: http://www.ich.org/cache/compo/1325-272-1.html#Efficacy). Clinical Data are initially summarized in detail in Module 2.7 and then, in a more integrative summary fashion, in Module 2.5.

The summaries contained in Module 2 build, as noted, on the information below. In considering the construction of Module 2, remember our adage from the first chapter — Begin with the End in Mind. Module 2 is the "End" and is built from the bottom up, but with an eye to its overall intent as a summary and "Sales" document. I use the term "Sales" document advisedly because I believe that is fundamentally what the document is. If the sponsor didn't think the product was worthy of getting to the market, the sponsor wouldn't go to the trouble of putting all this paperwork together. So, presumably, the sponsor believes the risk:benefit ratio for the product justifies moving forward. Assuming that is true, then the analysis and summarization of the data are means of convincing a regulatory agency that the risk:benefit ratio is heavily weighted in the *benefit* direction. Essentially then, Module 2 wages an argument, based on and supported by the data contained in the document, that sells this concept to the regulators.

As an example, consider the clinical information. Table 10-1 identifies the components of Module 5, the individual study reports, from each of the component phases of clinical development. Module 5, then, is comprised of sections for pharmacokinetic data, pharmacodynamic studies and the various types of later trials — Phase 2 and 3 studies. Note that Module 5.3.7 contains case report forms and patient listings — needless to say, there are lot of volumes in that module.

Table 10-1. NDA - Clinical Efficacy: Building from the Bottom Up

Module 5 (Individual Study Reports)	
Module 5.3.1	Reports of Biopharmaceutics Studies
Module 5.3.2	Reports of PK Studies using Human Biomaterials
Module 5.3.3	Reports of Human PK Studies
Module 5.3.4	Reports of Human Pharmacodynamic Studies
Module 5.3.5	Reports of Efficacy and Safety Studies • Controlled Studies • Uncontrolled Studies • Analyses of More Than One Study
Module 5.3.6	Post-Marketing Experience
Module 5.3.7	CRFs and Patient Listings

Then, moving to Module 2.7, the summarization process that takes place can be seen, building on the individual studies reported in Module 5. (See Table 10-2.)

Next, one more level up, what is "below" (Modules 5 and 2.7) can be seen to relate to the overviews in Module 2.5. As noted above, this is the section that presents the "sales" pitch. (See Table 10-3.)

Table 10-2. NDA - Clinical Efficacy: Building from the Bottom Up, Module 2.7

Module 2.7 (Summaries)	
Module 2.7.1	Summary of Biopharmaceutics Studies
Module 2.7.2	Summary of Clinical Pharmacology Studies
Module 2.7.3	Summary of Clinical Efficacy
Module 2.7.4	Summary of Clinical Safety • Exposure • Adverse Events • Laboratory Evaluations • Physical Findings • Special Groups, including drug interactions

Table 10-3. NDA - Clinical Efficacy: Building from the Bottom Up, Module 2.5

Module 2.5 (Overview)	
Module 2.5.1	Product Development Rationale
Module 2.5.2	Overview of Biopharmaceutics
Module 2.5.3	Overview of Clinical Pharmacology
Module 2.5.4	Overview of Efficacy
Module 2.5.5	Overview of Safety
Module 2.5.6	Benefits and Risks Conclusions

10.1.1 Translating the US NDA to the ICH CTD

For old timers such as I, who grew up with the old NDA format, we need some translation. Table 10-5 shows the old NDA sections on the left and their CTD relatives on the right.

10.1.2 Integrated Summaries of Efficacy and Safety

Beginning in 1985, long before the development of the CTD, all US New Drug Applications (NDAs; marketing applications) required an Integrated Summary of Safety (ISS) and an Integrated Summary of Efficacy (ISE); these summaries were to be integrated across all studies (non-human and human) to present an analysis of all the data enabling a "conclusion" regarding risk:benefit of the drug in the indication(s) being studied.

Implementation of CTD made these documents problematic — at least conceptually. Since the regulation wasn't changed, the answer to the question, "Do I still need to include the ISS and ISE in submissions now that the CTD format is required?" is still, "Yes!!" Table 10-4 places the ISS and ISE within the context of the CTD format.

Table 10-4. ISS and ISE Location within CTD Format

CTD Section	US Regulation	Comment
2.5 Clinical Overview (~30 pages)		
2.5.4 Overview of Efficacy	N / A	Not a US requirement; ICH guidance
2.5.5 Overview of Safety		
2.7 Clinical Summary (~50–400 pages)		
2.7.3 Summary of Clinical Efficacy	21 CFR 314.50(c)(2)(viii)	US requirements for Clinical Summary
2.7.4 Summary of Clinical Safety		
5.3 Clinical Study Reports		
5.3.5.3 Report of Analyses of Data from More than One Study (including Any Formal Integrated Analyses, Meta-Analyses and Bridging Analyses)	21 CFR 314.50(d)(5)(v) 21 CFR 314.50(d)(5)(vi)	ISE ISS

The key item regarding these documents is that they are analyses, not just summaries. Summaries can be quite "tabulative" in their nature and reach somewhat superficial conclusions. These documents are designed to be extremely analytical and thorough. There is more discussion regarding the FDA's philosophy about these documents on the FDA website. (http://www.fda.gov/cder/regulatory/ersr/ISS_ISE_clarification.htm) Although there aren't designated spots for them in the CTD format, the FDA has been pretty clear about where they go in a submission to the US. They should be included in 5.3.5.3. That section is entitled, "Reports of Analyses of Data from More than One Study (Including Any Formal Integrated Analyses, Meta-Analyses, and Bridging Analyses)".

Table 10-5. Comparison of US NDA and CTD Formats

Index of NDA (21CFR 312) or BLA (21CFR 314)		
1.	Index	Module 1
2.	Labeling (Draft / Final Printed Labeling)	Module 1
3.	Summary (21CFR 314.50 (c))	Module 2
4.	Chemistry Section	Module 3 Module 2.3
A.	Chemistry, Manufacturing, and Controls (21CFR 314.50 (d)(1))	
B.	Samples (21CFR 314.50 (e)(1))	
C.	Methods Validation Package (21CFR 314.50 (e)(2))	
5.	Non-Clinical Pharmacology and Toxicology (21CFR 314.50 (d)(2))	Module 4 Module 2.4 Module 2.6
6.	Human Pharmacokinetics and Bioavailability (21CFR 314.50 (d)(3))	Modules 5.3.1-4 Module 2.5 Module 2.7.1 Module 2.7.2 Module 2.7.5 Module 2.7.6

Table 10-5. Comparison of US NDA and CTD Formats (continued)

7.	Clinical Microbiology (21CFR 314.50 (d)(4))	Module 5 Module 2.5 Module 2.7
8.	Clinical Data Section (21CFR 314.50 (d)(5))	Module 5.3.5 Module 2.5 Modules 2.7.3-2.7.6
9.	Safety Update Report (21CFR 314.50 (d)(5)(vi)(b))	Module 5
10.	Statistical Section (21CFR 314.50 (d)(6))	Module 5
11.	Case Report Tabulations (21CFR 314.50 (f)(1))	Module 5.3.7
12.	Case Report Forms (21CFR 314.50 (f)(2))	Module 5.3.7
13.	Patent Information (claiming drug) (21USC 355 (b) or (c); FDA Form 3542)	Module 1
14.	Patent Certification (claiming drug) (21USC 355 (b)(2) or (j)(2)(A); FDA Form 3542)	Module 1
15.	Establishment Description (if applicable) (21CFR 600)	Module 1
16.	Debarment Certifications (FD&C Act 306 (k)(1))	Module 1
17.	Field Copy Certification (21CFR 314.50 (l)(3))	Module 1
18.	User Fee Cover Sheet (FDA Form 3397)	Module 1
19.	Financial Information (21CFR Part 54; FDA Forms 3454, 3455)	Module 1
20.	OTHER	Dependent on information

How do sponsors respond to the requirements to include ISS and ISE analyses in their submissions? See Table 10-6. The data come from the FDA. From 70 to 3,519 pages and 86 to 10,469 pages!! Those are some major differences.

Table 10-6. Page Variations in ISE and ISS Submission in US

Application A	
Section 2.7.3	70 pages
Section 2.7.4	86 pages
Application B	
Section 2.7.3	697 pages
Section 2.7.4	904 pages
Application C	
Section 2.7.3	3,519 pages
Section 2.7.4	10,469 pages

Source: Gensinger, FDA, 2006

10.2 Individual Patient / Study Data Flow

Figure 10-4 illustrates the flow of data from a protocol and a patient through to a clinical study report. Next, we will look at the relationships between the clinical study report, the statistical analysis plan (SAP), the process of data management, and case report forms (CRFs). As usual, we will be starting from the end and working backward, so we will begin with the Clinical Study Report.

10.2.1 Clinical Study Report

The clinical study report (abbreviated CSR) is the end product of a clinical study. So, in keeping with the philosophy of "Begin with the End in Mind", we will start here. Two important guidances for your consideration are the ICH E3, "Structure and Content of Clinical Study Reports," (www.emea.europa.eu/pdfs/human/ich/013795en.pdf) and ICH E9, "Statistical Principles of Clinical Trials" (www.emea.europa.eu/pdfs/

human/ich/036396en.pdf). The ICH E3 guideline provides all the details for basic formatting of the document (including numbering schemes; text fonts and sizes, page margins, header and footer composition, and page numbering. See Table 10-7.).

The goal of a study report is to produce a well-organized and unambiguous review that fully integrates the clinical and statistical elements of the study into one coherent report. The data from the clinical trial, represented by the CSR, contribute additional information to the target product profile (product labeling) developed at the beginning of the project.

There are several important elements included in the CSR. The CSR must first relate how the study was actually conducted. The protocol was the guideline for the intended conduct of the study, but the CSR reports what was actually done. The CSR then describes, as unambiguously as possible, the results of the study. Conclusions from the results are then drawn and woven into a discussion of the product's overall safety and effectiveness. If it is an early study, much of the contextual information is provided from non-clinical studies. As more and more clinical data are obtained, the information can be integrated into a broader picture of the compound's safety and effectiveness.

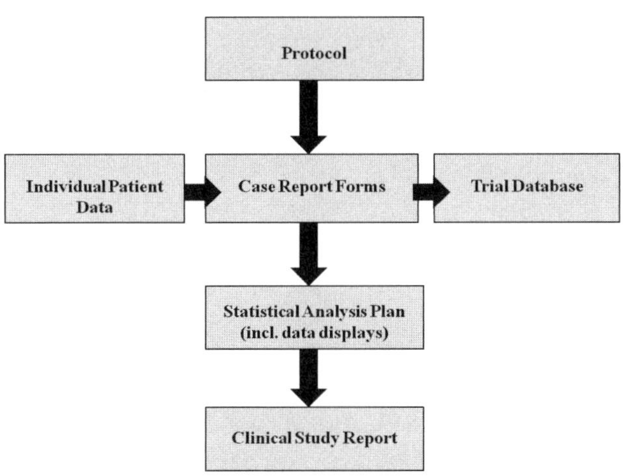

Figure 10.4. Clinical Trial Data Flow

Table 10-7. Clinical Study Report Standard Format

1.	Title Page
2.	Study Synopsis
3.	Table of Contents
4.	List of Abbreviations
5.	Ethics Section
6.	Investigators and Study Administrative Structure
7.	Introduction
8.	Study Objectives
9.	Investigational Plan
10.	Study Patients
11.	Efficacy Evaluation
12.	Safety Evaluation
13.	Discussion and Overall Conclusions
14.	Tables, Figures, Graphs
15.	Reference List
16.	Appendices

By the way, I believe there should be a substantial integrative section in the CSR (Section 13: Discussion and Overall Conclusions). There are some, including some very large companies, who believe that the study itself should simply be reported and other documents will provide the integrative function — different philosophies.

The study synopsis (*Section 2* of the CSR) deserves a couple of extra words. One should be able to lift the synopsis from the study report and have a very good idea what happened in the protocol. That doesn't mean it is an endless recapitulation of the rest of the report, but the key information and results need to be in this section.

Section 5 is an ethics section that confirms the protocol, and the amendments, were reviewed by IRBs or ethical committees, and that the protocol was conducted according the principles of the Declaration of Helsinki. The procedure and timing for obtaining the informed consent is also described in this section. While relatively small sections in the study report, from our prior discussions, it is obvious how important they are.

Section 6 identifies the investigators participating in the study and the study's administrative structure. Essentially, all individuals and groups responsible for key elements of the study, its conduct and its reporting, are identified in this portion of the CSR.

Steering Committees, if any, used to any purpose during the conduct of the study, are also identified in Section 6. Steering committees (or monitoring committees) can have any of several functions. For example, an internal committee (i.e., composed of employees of the sponsor or consultants hired by the sponsor) might be created to review safety data for clinical programs. This would include reviewing data from different studies for the same product or different studies for different products. If such a committee is in place and data from this particular study were reviewed by the committee, their function and the processes employed in reviewing the data (e.g., blinded or not, summary data or individual patient data) would be identified in this section.

Data Monitoring Committees (DMCs) can also have several forms. For example, the committee might have been constituted to review safety data or efficacy data or both. Whatever its function, the level of independence from the sponsor and the extent of data access for the DMC need to be identified. In this section are also listed the institutions, statisticians, laboratory facilities, and contract research organizations used. (See discussion regarding DMCs in Chapter 12.)

Section 7 is designated an Introduction. The Introduction to the study report needs to put this study into the context of the overall development of the product and describes how this protocol fits into the overall development plan.

A key question answered in the Introduction is: "What is this study contributing to the development plan that justifies conducting the study as described in the protocol?" It is not good enough to just say, "Well, we needed to do one more study to demonstrate effectiveness for this particular product or to generate maybe a little bit of extra safety data." The study report should clearly indicate the answer to the question. In reality, the Introduction should need to only recapitulate information in the protocol's introduction, because the justification should have been convincingly presented there. It is also important to identify any relevant meetings and agreements with regulatory authorities regarding the overall development program and the study in particular.

Section 8 in the CSR explains the study objectives. In general, those should be clearly and simply identified in the protocol, so recapitulation is all that is needed. The exception to this might be some change in the protocol, particularly if something were occasioned by a decision by the monitoring committee.

Section 9 of the Study Report describes the overall investigational plan in terms of design, controls, and analytical and statistical methods.

Section 9 describes the overall study design and plan, including the use of control groups, identification of the study population by inclusion and exclusion criteria, the treatments employed (doses, blinding, and compliance achieved), the efficacy and safety variables assessed, a description of data quality assurance, the statistical methods employed (describing the sample size and the analyses conducted), and a description (and discussion) of any deviations from the original plans.

This is an important section of the CSR. The analyses discussed need to be correlated to the Statistical Analysis Plan. Any deviations from plan detailed in the original protocol and / or the original Statistical Analysis Plan — and what was actually done in conduct of the protocol and / or in conduct of the data analyses — need to be clearly explained and justified.

Section 10 focuses on the individual patients. In particular, we want to know who got into the protocol and what actually happened to them during the protocol. How many enrolled? How many dropped out?

Different analytical populations are addressed here. For example, how many fit into the "**safety population**" (generally anyone who got a drug or placebo) or the "**efficacy population**" (generally anyone who got a drug or placebo and has some kind of data)? How many completed all aspects of the protocol as written ("**per protocol population**")?

This section also identifies whatever **protocol deviations** were identified. As a normal course of events in the conduct of a study, protocol deviations will be permitted. For example, the enrollment criteria specify an age range from 18–70 years of age. An investigator identifies a patient who otherwise meets all enrollment criteria, but is 71 years of age. It is likely that this patient will be permitted to enter the study by the medical monitor. Nevertheless, the deviation will need to be identified in this section of the report. Hopefully, there won't be deviations (protocol violations) that were not formally approved. But, if there were, they need to be reported here.

Demographic characteristics are also identified (usually in tabular format) in this section. For example, in a study evaluating an agent to treat congestive heart failure, it is important to understand for each treatment group how many patients had different stages of heart failure, how many had a prior history of diabetes, how many had a prior history of hypertension, how many had a prior myocardial infarction, etc. The demographic data are important for simply putting the results in context — for example, did patients who had heart failure and a prior myocardial infarction fare better with the treatment than did patients who had not had a prior myocardial infarction? — or to understand whether there was (presumably unintentionally) some kind of systematic alteration in allocation of patients to treatment groups that resulted in differential efficacy responses.

Reviewing and assessing such data also provide additional understanding of differential safety responses (e.g., adverse events or drop-out rates) between treatment groups. Treatment compliance is usually reviewed and analyzed in this section.

Section 11 of the CSR covers evaluation of efficacy. Each of the specified data sets (for example, "efficacy population" and "per protocol population") is reviewed and analyzed in terms of meeting the specified statistical endpoints. Subgroups, identified, for example, by baseline characteristics and / or demographics, are compared in terms of responses. It is important to clearly state whether these subgroup analyses apply to groups identified prior to the start of the study and the analyses were specified in the Statistical Analysis Plan and the Protocol or whether the subgroups were identified post hoc (that is, after the data from the study were available).

Other issues are also addressed in Section 11. What were the drug-dose, drug-drug, and drug / dose-disease interactions? Were there any center effects noted? Were there subgroups identified? What happened in the analyses of these patients and groups? Were there differences from the main population? These are the type of questions addressed in this section.

I believe it is important to draw conclusions from these data analyses. It is important to note, however, that different individuals and different companies have different viewpoints about "drawing conclusions" in this Section. I personally believe it is important, but I recognize there are different points of view.

Section 12 reviews safety within essentially the same types of context as the review of efficacy in Section 11. All of the assessments are framed within the extent of exposure. Exposure is usually analyzed from at least three points of view — from dosing (e.g., 50 mg versus 150 mg versus 300 mg), duration (e.g., 2 weeks versus 4 weeks), and cumulative dosing (daily dose times the duration). In particular, adverse events, other serious adverse events (including deaths), clinical laboratory assessments, vital signs and various physical findings are reviewed and analyzed within the context of exposure. (Although addressed in the safety section, extent of exposure is also a relevant contextual analysis for efficacy as well.) Again, in my opinion, conclusions should be drawn based on the analyses presented.

These two sections, the efficacy review in Section 11 and the safety review in Section 12, are extremely important portions of the report. In the context of beginning with the end in mind, consider the output into these two sections when the protocol is initially drafted and the Statistical Analysis Plan developed — before the study ever starts.

Section 13 is the portion of the report where the information is integrated and we can make some (hopefully intelligent) comments about the risk:benefit ratio for the compound in light of the data generated during this study. I believe — again, philosophies differ on this — it makes sense to also put data from this study into the context of other, previously-conducted studies' data.

Section 14 is the portion of the report that will contain all the tables and graphs developed in the context of the statistical analysis plan. *Section 15* lists all articles or studies referred to in the CSR. Generally, these are studies cited in the Introduction, Study Objectives and Investigational Plan parts of the CSR (and, usually, these are carried over from citations in the protocol).

I will warn you that *Section 16* is actually the largest of the whole report and will carry on for at least hundreds, and usually, thousands of pages, since it contains documents associated with the trial as well as patient listings.

10.2.2 Statistical Analysis Plan

Moving upward on Figure 10-4 (or backward chronologically) from the CSR, we will now consider the Statistical Analysis Plan. (This is also called the Clinical Analysis Plan.) What is the **Statistical Analysis Plan** (SAP)? It's a complete set of instructions designed to establish that the clinical data collection, the reporting procedures, and the analytical processes are all defined and very clear prior to the time that we start the study. As noted above, a good reference for developing the SAP is the ICH E9, "Statistical Principles for Clinical Trials" (www.emea.europa.eu/pdfs/human/ich/036396en.pdf).

The SAP is a formally-written document. It describes the analysis of each and every data point collected during the course of the trial. It particularly addresses how the elements critical to our analysis are identified and handled. It will also address the data auditing process, identifying which parameters and which data will be reviewed and audited 100% versus other data that may be less rigorously audited.

The SAP describes the detailed study objectives and the variables measured in the study that serve those objectives. The details of the planned statistical analysis of both the primary and the secondary variables are also described. In addition, mock-ups of all tables, figures and the data listings planned for the Clinical Study Report are included in the analysis plan as appendices.

There should be a very tight relationship between the processes described in the SAP and the analytical and data handling processes described in the clinical protocol. As noted above, any deviations from the plans to the execution of the study need to be clearly addressed in the CSR.

10.2.2.1 Strategy Meeting

The SAP itself is generally authored by the statistician attached to the study. However, I believe strongly that the decisions made that ultimately result in the approaches described in the document should derive from a single or several strategy meetings.

We discussed previously planning meetings associated with creation of the overall development plan. In a very real sense, the strategy meetings associated with a single protocol are a continuation of the development plan meetings. As for those meetings, there are several critical attendees at the protocol strategy meetings. These would include the project study director, a physician (who will be the medical monitor), a statistician, the clinical trial manager, the project manager, and the regulatory representative. Those are folks who are the minimum attendees at this particular meeting. You may also want to have other individuals from your data management group. This would be true, particularly if you happen to be large enough to have someone specifically focused on case report form design, data entry, data verification, database development, safety, etc.

Of course, it is possible that some of the individuals may be consultants. To the extent consultants fulfill critical functions associated with clinical trial development, data management, and regulatory strategy, their attendance is extremely important.

The goals of the meeting are to define the study objective(s), each of the variables that will be collected during the course of the study and the overall study design. Each of the analyses for the primary and secondary variables should be developed in detail and agreed upon.

10.2.2.2 Data Displays — Standard and Study-Specific

At this stage, it is also important to mock up the data displays. I believe it is very important to have this detailed discussion at this point, so it is clear where data collected in the trial will go, and to identify issues regarding analysis and display if there are any. Information developed during this meeting is then forwarded to the programmers who will provide feedback and pose questions if necessary. This is consistent with our focus on "Keeping the End in Mind".

At the conclusion of the interactive discussions, all of the standard data presentations reflecting depictions of demography, safety tables, figures, and efficacy will be created and programmed.

The first time an organization does this, it is the hardest. Subsequent protocols should be able to build on the examples from these studies and undergo, perhaps, minimal modification. Additional requirements, of course, may arise in the development program.

The mock-ups for the presentation should be standardized as much as possible, including the titles, headers, footers, fonts, etc. In addition, displays related to variables, sorting of variables (e.g., by investigator, treatment group, patient, etc.), and analyses should be similar from table to table, figure to figure and listing to listing. This facilitates review by the person(s) writing the CSR and anyone (e.g., regulators) subsequently reviewing the data. The monotony of locating similar items in the same places will make the information much easier to review.

The goal is to modularize as much of this as possible, so data collection processes, including case report forms, and statistical analyses can be relatively standardized across the development program. As much as possible, these are designed once and then used for all studies. That being said, reality rears its ugly head frequently in the course of studies and programs, and some tweaks must be made to the CRF pages and to the data displays. However, through thoughtful preparation, the "tweaks" can be minimized. In addition, there will always be study-specific data presentations, particularly for studies designed to ask a very specific question or set of questions. Each collection and display vehicle should be developed with an eye to the eventual data presentation in regulatory submission packages.

Mock-ups of the primary and secondary endpoints should be broken down by respective populations of interest. For example, the **evaluable population** (also called efficacy population; subjects who were treated with at least one post-dose efficacy assessment), the **safety population** (subjects who were treated, even with a single dose), the **intent-to-treat population** (all subjects who were randomized, whether they received treatment or not), the **per-protocol population** (all subjects who completed all aspects of the protocol, including follow-up), etc. In addition, subgroups of interest defined at the beginning of the study will also be analyzed and mock-ups of the primary and secondary endpoints for the subgroups should also be designed.

Detailed mock-ups should also be designed for each of the analytical frameworks we will apply to efficacy or safety data. Examples would include trends by time, by center / site, by country (in multinational studies), by defined subgroups, etc.

It is useful to identify **subgroup analyses** during the planning phase so that stratifications can be performed in the randomization process in order to facilitate balancing of groups. However, it is also possible that subgroups will also only become evident once you have looked at the actual data from the trial. Those types of subgroup analyses are quite useful in terms of understanding your product. However, keep in mind that if the analyses weren't pre-determined and weren't conducted without knowing the data, they will only serve informational purposes. Many companies will assume those analyses will now be approvable endpoints.

They can be re-studied in a new study that appropriately controls for them, but this study won't do it!

Center and site analyses are important to ensure that a center or even a group of centers significantly (in a statistical sense) change the direction of the responses in the data. For example, Center One and Center Two each contributed 25 patients to your study. However, the study only enrolled 75 people and there were 10 OTHER centers participating in the study. There is a clear weighting bias in this particular study. As you analyze the efficacy data, the responses will need to be compared across centers to ensure poolability of the data. While a bit contrived as an example, you understand the point. These types of effects need to be tested across countries also, when conducting multi-national studies.

10.2.2.3 Data Presentation Programming

I believe — very strongly actually — that the programming required for the tables and figure shells needs to be put in place *prior to the entry of the first patient into the protocol.* Ideally, the shells have been tested with dummy data, but if not, they should be tested with the data (keeping responses blinded) from the first few patients. The number is dependent upon the total required for the study. If 1,000 people are being enrolled, you might wait for the first 15–25 patients to test the shells. On the other hand, if 100 people are entering the study, you want to test the shells with data from, perhaps, five people. It is not unusual for the tables and figures to require tweaking. This process needs to be accomplished in such a way that the data from these patients are not compromised.

One final item in the discussion of the analytical plan is that there should be an "official" meeting — official only in the sense that people actually do sit down together and produce a set of minutes — to ensure concordance between the protocol, the Case Report Form, the tables, listings and figures, as well as the database. If you can do that up front, you will have gone a long way to solve potential issues and prevent a lot of data management problems.

10.2.2.4 Data Presentation Examples

In this section we will present several highly schematized examples of what some of these tables and listings might look like.

Table 10-8. Subject Completion Summary

Sponsor									
Test Article									
Report No.									
Treatment		**Screen**	**Enroll**	**Random**	**Visit 1**	**Visit 2**	**Visit 3**	**Visit 4**	**FU**
Test		N (%)	N (%)	N (%)	N (%)	N (%)	N (%)	N (%)	N (%)
Control		N (%)	N (%)	N (%)	N (%)	N (%)	N (%)	N (%)	N (%)
Total		N (%)	N (%)	N (%)	N (%)	N (%)	N (%)	N (%)	N (%)

Our first example, Table 10-8, describes study disposition, the number of subjects enrolled (that is, signed a consent), randomized (hopefully, the same number), those who completed specific visits, etc. This type of presentation shows the time course of patients as they evolved through the course of study. There are other, equally valid means of representing these types of data.

The next example is a Summary of Subject Status (Table 10-9) and might look at further analysis patient participation.

In this case, the focus is on the number of patients completing the protocol as prescribed. The control group and the test group are identified. The actual numbers — as well as the respective percentages — for each group are represented. There are many potential variations of this type of table.

This type of table can be particularly challenging if there are multiple treatment groups in the trial. For example, treatment groups might include placebo, the low dose, the medium dose, and the high dose of our product, and even an active comparator. The display must include, in an easily-decipherable format, the appropriate information.

How many people had adverse events? How many people were lost to follow-up? How many did you, the sponsor, terminate because of protocol violations? How many patients withdrew their consent once they were in the middle of the protocol because it was too inconvenient or they didn't like the side effects? Whatever the issues are, all of that needs to go in here.

Essentially, you are providing a complete scorecard of what happened to all of the subjects.

Table 10-9. Summary of Subject Status

Sponsor Test Article Report No.						
		Control		**Test**		**All Groups**
No (%) of Subjects		N (%)		N (%)		N (%)
Protocol Completion		N (%)		N (%)		N (%)
Non-Fatal Adverse Event		N (%)		N (%)		N (%)
Lost to Follow-Up		N (%)		N (%)		N (%)
Terminated by Sponsor		N (%)		N (%)		N (%)
Withdrawal of Consent		N (%)		N (%)		N (%)
Death		N (%)		N (%)		N (%)
Other		N (%)		N (%)		N (%)
Total		N (%)		N (%)		N (%)

Table 10-10 is a means of portraying demographic analyses. The control and the test groups are separated and compared, respectively, by sex, race, age, height, weight, etc. This type of representation can be expanded to cover any demographic items of interest.

Table 10-10. Demographic Summary by Treatment Group

Sponsor **Test Article** **Report No.**								
				Control		**Test**		**All Groups**
Sex N (%)		Female		N (%)		N (%)		N (%)
		Male		N (%)		N (%)		N (%)
Race N (%)		White		N (%)		N (%)		N (%)
		Black		N (%)		N (%)		N (%)
		Hispanic		N (%)		N (%)		N (%)
		Amer Ind		N (%)		N (%)		N (%)
		Other		N (%)		N (%)		N (%)
Age (years)		Mean (+/- SD)		xx.x (xx.x)		xx.x (xx.x)		xx.x (xx.x)
		Median		xx.x (xx.x)		xx.x (xx.x)		xx.x (xx.x)
		Range		xx.x (xx.x)		xx.x (xx.x)		xx.x (xx.x)

For example, in a study of coronary artery disease, we might want to know how many people entered the study with a history of diabetes, or hypertension, or coronary artery disease. Those data can be represented on this type of table.

Longer studies or those with sicker patient populations are likely to include patients who will die during the course of the study. The representation of those patients is in a listing of deaths. (See Table 10-11.)

Table 10-11. Listing of Deaths

Sponsor								
Test Article								
Report No.								
Subject ID		**Adverse Event**		**Treatment Group**	**Date First Rx**	**Date Last Rx**	**Date of Death**	**Causality**
		Preferred Term	**Verbatim Term**					
006 244		Myocardial infarction	Heart attack	Placebo	11 Nov 2009	11 Dec 2009	15 Dec 2009	Unrelated
061 007		Etc.						
061 121								
101 144								

A similarly formatted listing would represent the adverse events occurring by treatment group in the study. In that case the listing would need to include both the verbatim term (that term used by the investigator or investigative site to represent the event) and the preferred term (that is a term facilitating the coding and analysis of data; it is determined by the "dictionary" your data management people use). An important feature of the death and adverse event listings is the designation of relatedness (causality) of the treatment to the event. In all cases, the investigator needs to record their opinion. The sponsor might offer a different opinion (and perhaps it could be well-reasoned), but the one that will count for tabulation purposes is the one provided by the investigator.

Laboratory results are usually depicted in both listings and tables, and frequently figures as well. Table 10-12 is an example of a tabular representation of laboratory data.

Table 10-12. Summary of Laboratory Results — Hematology

Sponsor **Test Article** **Report No.**								
		Baseline	Visit 1	Visit 2		Baseline	Visit 1	Visit 2
WBC (Normal Range, units)		x.x	x.x	x.x		x.x	x.x	x.x
Mean		x.x	x.x	x.x		x.x	x.x	x.x
Mean Change from Baseline		x.x	x.x	x.x		x.x	x.x	x.x
Median		x.x	x.x	x.x		x.x	x.x	x.x
Median Change from Baseline		x.x	x.x	x.x		x.x	x.x	x.x
Range		X - Y	X	Y		X	Y	X
% Abnormal		%	%	%		%	%	%

Whatever format is chosen should work for essentially all lab data. In this case, we picked hematology data as an example. To better understand potential issues with laboratory results and safety, the data should be evaluated (and displayed) in several different ways.

Table 10-12 is a more representational format. The mean, the median, and changes from those respective values over time are represented. Although not depicted here, standard deviations and standard errors, as appropriate for the data, can also be represented.

Next, you will want to look at those lab data in a trend format. (See Table 10-13.)

Table 10-13 is one way to represent trends by visit (see the title), by treatment group (control vs. test), by variable (WBC, RBC, etc.). As noted above, multiple treatment groups can complicate these data representations considerably.

Table 10-13. Summary of Laboratory Trends — Hematology
Baseline — Visit 1

Sponsor **Test Article** **Report No.**								
		Control (Trend from Baseline)				**Test (Trend from Baseline)**		
Variables	N	**% Down**	**% No Change**	**% Up**	**N**	**% Down**	**% No Change**	**% Up**
WBC	N	%	%	%		N	%	%
RBC	N	%	%	%		N	%	%
Hgb	N	%	%	%		N	%	%
Hct	N	%	%	%		N	%	%
Platelets	N	%	%	%		N	%	%
Eosinophils	N	%	%	%		N	%	%
Basophils	N	%	%	%		N	%	%

Another way of looking at the lab data is represented in Table 10-14. This format is generally referred to as a change table. In this case, the question addressed by this representation is, "How many of the subjects had low values, normal values, high values, for each laboratory parameter at baseline?" Then, "Given where they started, what happened to those parameters during the study?"

For example, Table 10-14 illustrates the data at Visit 1 (for example, at Week 2), compared to baseline. Another table would represent data from Week 4 compared to baseline. Another table would represent Week 6 compared to baseline, and so forth. All of that information is assembled into a format that enables an understanding of temporal change of a parameter relative to its starting point.

The goal of these (and, indeed, all) of the presentations is to enable better understanding of your product and its use in the target population across time.

Table 10-14. Summary of Laboratory Value Changes — Hematology Baseline — Visit 1

Sponsor								
Test Article								
Report No.								
Variable		**Control (Trend from Baseline)**				**Test (Trend from Baseline)**		
	BL	**Lo**	**N**	**Hi**	**BL**	**Lo**	**N**	**Hi**
	Lo	%	%	%	Lo	N	%	%
WBC	**N**	%	%	%	N	N	%	%
	Hi	%	%	%	Hi	N	%	%
	BL	**Lo**	**N**	**Hi**	**BL**	**Lo**	**N**	**Hi**
	Lo	%	%	%	Lo	N	%	%
RBC	**N**	%	%	%	N	N	%	%
	Hi	%	%	%	Hi	N	%	%

There are many other analyses and representations than can, and should, be done to evaluate the data generated in clinical trials. The full scope of them is beyond the intent of this book. Hopefully, however, you have some inkling of how to consider the data questions of interest to you.

10.3 Data Management

Moving further up Figure 10-4, we will next consider individual patient data, the case report form and the trial database. Clinical data management constitutes the set of activities associated with the data generated in clinical trials. From the first entry of data into a case report form through to the production of the tables, listings, and figures representing the data captured during the conduct of the study, data management serves as a crucial interface between the clinical trial protocol and the final output in the Clinical Study Report.

In data management, one of the key issues is the quality of the clinical trial data. The quality of the data plays a major role in determining whether the trial and, ultimately, the product, is a success or failure. Given its importance, management of data requires a pre-planned strategy. As a general statement, data flow and handling must conform to established standards for collection, processing, analyzing and reporting the data.

Keeping the End in Mind, we must understand (and manage) the flow of data from the initial case report form to the final study report, and then to the various sections of the Common Technical Document (CTD) in order to obtain marketing approval for our product.

However data management is done in your company, whether internally in a fully-staffed department or externally in a fully-outsourced activity to a contract organization or as an operational hybrid between the two, formal, written, internal standard operating procedures (SOPs) are required to describe the process by which data will be collected, processed, managed and reported. These SOPs essentially serve as the "recipes" to be followed for managing data.

The data management plan itself is cooperatively created through interactions of (at least) medical, clinical, biostatistics, programming, medical writing, regulatory, and data management personnel and describes in detail the role of data management as a function in the process of data flow.

10.3.1 Case Report Form (CRF) Design

It is important to remember the connection between the CRF and the database. CRFs can be paper or electronic. Regardless of the format, it is crucial that there be a seamless interface between the data entry onto the CRF and the data entry into the study database. Design of the CRF and the screens for data entry should be done in tandem to facilitate accurate and efficient incorporation of data into the database. An early step in this process is annotation of the CRF, so that all items on the CRF can be represented in the database. Each item (variable) is assigned a name and then is characterized by the type of data (number, alphanumeric, date, etc.), the length of the field, and the type of coding involved.

As a general rule, the computer screens for data entry should mirror the look and feel of the CRFs as much as possible. Doing that reduces the errors at the point of data entry significantly. Currently, however, many studies are conducted with some form of direct data entry. For example, electronic data capture involves direct data entry by the investigative site into the computer rather than onto paper and the data (with appropriate edit checks, e.g., you can't enter a name into a date field) go directly to the database for review by the data management team. An alternative methodology involves data faxing in which the CRF page is coded in such a way that when faxed, the data go to the database (again in the presence of appropriate edit checks). It is of interest that patient diaries are also moving to fully electronic formats as well; so far there have been some challenges, particularly in groups of individuals (e.g., elderly) or countries that don't have much experience with electronic devices.

To facilitate data entry, a **CRF Completion Guide** should be provided to the investigative sites. At investigator meetings, typically there will be a session in which each unique CRF page is identified and the requirements for completing it are reviewed with the investigators and study coordinators.

10.3.2 Data Validation

Following testing of the database with data entry and data output (tables, listings, figures), the next step in the data management process is data validation. This includes checking and editing of data. The SOPs mentioned above will detail specific procedures for performing these processes. (Or, if this process is outsourced, verify that your vendor has SOPs to cover this step.)

Data entry will be followed by data verification. Each of these components is an essential element of the overall data management process.

10.3.3 Query Generation and Resolution

Once data are entered into the data base and verified that they are accurate (i.e., they are in the right format and the data reflect entries on the CRF pages), the next phase includes the generation of queries. **Queries** are necessary in those cases in which the data are illegible, there are conflicting responses to questions, the sequences of dates don't make sense, etc. A whole host of things can generate queries. Ultimately, each one generated needs to be resolved in some formal manner. Either it is changed or fixed or actively ignored. The resolution (even if the resolution is "no change") needs to be documented.

Expanding on the process a bit, as you enter the data into the computer, edit checks are built in. An **edit check** will define ranges or lists of entries that are acceptable (and, in turn, those that are not acceptable). For example, an edit check might specify that the numbers entered into the blood pressure blanks might need to be above "10 or 20", for example, but couldn't be "1,000". The specifications are built into the edit checks ensure data are useful.

If you look at what's on the Case Report Form and enter the data, the edit check might flag the response indicating, for example, that the entry is outside the acceptable range. Then, a query is sent to the investigative site. Currently, most of the initial queries are processed by email and documented electronically. The CRA will be copied on the query and will, then, need to ensure that the site had adequately responded to the query.

It is entirely possible that the data are accurate; in that case, an exception is noted in the specific field in the database. Frequently, the errors are straightforward, such as missing decimal points or transcriptional mistakes. Query generation and resolution assists in maintaining the integrity of the database.

Queries are important for at least two reasons. First, they enable verification (or correction) of information that will go into the database (this is good) and secondly, they require (in aggregate) huge blocks of time (this is bad). Ideally, systems (including instruction and training) are put in place to produce high quality data, so fewer queries are generated. On the other hand, when queries are generated, they need to be resolved in the most efficient manner possible.

10.3.4 Pre-Study Report Data Analysis

Prior to the time the study report is actually assembled, the data are reviewed initially in a blinded fashion in an attempt to review the data from a generalizability perspective. The point is not to develop conclusions regarding the data; rather, to understand their presentation and understand potential conclusions from looking at the data (without, at this point, making assumptions about the identity of Group A, or B, or C and so forth).

As an example, data could be reviewed to assure adequate randomization; based on this look, might there be some issues in comparing or contrasting the Groups? By looking at the data in a number of different ways (before the blind is broken), potential avenues of questioning (or data analysis) are generated; these will assist the understanding of the data and ultimately, the preparation of the study report.

I think this is an important step. Many folks head right to the unblinded data and then determine there are issues with presentation. Discovery at that point is associated with its own sets of issues.

In the Pre-Study Report Data Analysis, other issues that will plague the data analysis process can also be reviewed. Although many of these issues can (and should) be anticipated at the time the SAP is prepared, in fact, some issues will only be identified at the time real data are evaluated.

For example, what happens if the data are missing? How will they be handled? Which ones are really crucial? What data are so far out there (outliers) that validity should be questioned? What about data transformations? (Data transformation entails taking an element from one or two or three different sections in the case report form and performing a calculation from those data to arrive at another data point.) The calculations, of course, must be verified. Other tasks that will be performed on the real (but still blinded) data will be resolution of questions regarding coding of adverse events and concomitant medications. This is important since automatic coding dictionaries, while very good, are still not perfect, and inconsistencies should be resolved prior to unblinding.

Subgroup and secondary analyses should also be reviewed to understand their representation in the final data presentations.

Additionally, what data might be considered for exclusion from analysis? I say "considered" simply from the standpoint that a regulatory agency may, at the end, throw it back in. The ICH E9 Guidance specifically addresses data exclusions by saying the data may be excluded only if:

- the entry criteria were measured prior to randomization;
- the determination of eligibility violations can be made objectively (i.e., before knowing how the study came out);
- all subjects are equally scrutinized for eligibility violations; and
- all detected violations of the particular criterion are excluded.

Therefore, the sponsor can't pick and choose which subjects might be included and which people might be excluded, or which data might be included or not excluded. The conclusion: Exclusions may occur only for valid statistical reasons.

10.4 Case Report Forms

Next, we turn to CRFs.

The purpose of the Case Report Form is to capture *relevant* data for presentation and analysis of results. I emphasize the word "relevant" because many times, in the initial discussions, a question is asked, "Gee,

I wonder what would happen if we collected that?" The question could relate to any one of 8,000 things that could be collected in a study. My recommendation, however, is to *collect only the information that is relevant for this particular study*. Frequently, all those extra "what if" questions will simply weigh the study down and yield an inconclusive result.

The information collected in the Case Report Form importantly links the patient to the protocol and, in turn, to the database and from there to the marketing application.

As noted above, in the discussion of data displays, it is important to standardize as much of the formatting involved in the CRFs as possible. This starts with the basic page formatting — headers, footers, fonts, etc. For example, the locations for subject identification, dates, laboratory data, etc. should be, as much as possible, identical from one form to the next. That makes the forms easier for the sites and the data entry people to use. It includes paralleling aspects of the protocol, such as the check lists representing inclusion and exclusion criteria. Also, as noted above, the layout of the CRFs should correspond as closely as possible to the layout of the data entry screens.

10.4.1 Helpful Hints

These comments are based on years of experience and more mistakes than I care to remember!

The directions for use of the Case Report Forms should be clear and concise; this includes directions on the CRF pages themselves as well as in the CRF Completion Guideline document.

As general guidance, I would suggest complex questions be divided into several smaller questions. A means of collecting information in this circumstance is to break the question into a subset of questions requiring individual answers; in general, the information is clearer. Avoid general comment spaces "like the plague". Don't put in blank spaces and just ask people to provide comments. Unfortunately, they will; then one is stuck figuring out how to deal with them in the database. The exception to this is on pages provided for Adverse Event and Serious

Adverse Event reporting. You want comments and as much explanation as necessary on those pages.

A useful technique is to anticipate most of the responses and create a format in the Case Report Form in either a question box format (to check "yes" or "no") or in a format that requires a specific value be entered.

A very important guidance is: *Ask a specific question only once.* For example, on one page of the CRF, the patient's age is recorded. On another page, the patient's birth date is recorded. You might be surprised how often the two don't match up. So, ask only one (in this case the birth date is usually the best bet.)

Also, repeat the mantra: *Check boxes are good and text spaces are bad.*

I would suggest arranging predetermined responses in a meaningful order and then employ that arrangement consistently. For example, if the question "How is your pain today?" is being quantified, the CRF might always start with the low number on the end on the left side and the high number on the right side, or vice versa, whatever is intuitively logical, but do it consistently. Likewise, arrange the assessment in some sort of meaningful fashion such that the person who looks at it can assess it and put the response into the proper location.

Date and time fields on Case Report Forms are always interesting to say the least. Consider the effects of date changes. For example, if a particular therapy might continue past midnight (a common problem in studies done in intensive care units), this is an issue. Use of an antibiotic, for example, or a pain medication, or some other intravenous form of therapy, might occur over a time period which spans a couple of dates. Perhaps therapy begins at 11 o'clock at night on one night and continues through to 1 o'clock in the morning the next morning. On which day did that therapy occur? An accurate handling of such information should be included in the Case Report Form and accurately reflected in the database.

Then there is the simple issue of date formatting. You are likely to know that representations of dates in essentially all countries of the world are different than in the US. Our June 7, 2010 might be written 6-7-2010. A European would look at that shorthand and be convinced you mean July 6th. The day-month-year format, with the month being represented as a three-letter abbreviation followed by a four-digit year, seems to be pretty easily understood and implemented even in the US. This format would be 07 JUN 2010 for the date we meant to indicate.

The CRF book is generally formatted by visit (e.g., Screening, Baseline, Visit 1, Visit 2, Follow-up, etc.) with each of the requisite forms in the appropriate visit. Some forms will appear multiple times because they will be utilized at multiple visits. The coding at the top will allow discrimination from Visit to Visit.

The Case Report Form will also contain various log forms. When the log forms are created, particularly forms such as medication logs or adverse event form logs, it is best that the formatting is consistent from one portion of the study to the next portion of the study.

The variation in treatment regimens in the study (especially in intensive care settings) may present challenges in terms of the ability to record a dosage, dosage times, etc. Together with the persons responsible for data entry and data management, the format of these logs needs to be carefully considered in advance as best as possible — considering not only the present study, but the possibility of consolidating data across multiple studies.

Key presentations of adverse event data in the ultimate marketing application need to be clearly understood. However those data are to be presented, the collection of the data in the same format will facilitate processing of the data.

In general, the Case Report Form design ought to permit transcription directly from source documents. One thing that facilitates data transfer (and verification) is for the data boxes to be located on the right side of the CRF paper, so that the source document can be literally right next to the appropriate data box. Thus, the person doing the transfer has to travel a minimum distance to copy the information from the source document to the Case Report Form.

Whatever the units used need to be very clear; in almost all circumstances, units are preprinted on the CRFs or are included in a check box or might be represented by a code.

I think it is important to try to keep the investigative sites from having to perform any calculations or perform data transformation. Data transformations should be done by the in-house computer (for example, transforming a birth date to an age-at-study-entry data point). On occasion, however, the investigator / coordinator needs to perform a calculation, for example administration of a dose on a weight or body surface area basis. If this is necessary, provide a series of boxes to capture each step of the calculation, so that errors can be easily identified and rectified (if possible). The subject's weight would be recorded. The dosage per weight would be either preprinted or a box would be available for recording it. The final calculation would then be entered into the appropriate boxes on the CRF.

In addition it's also important if, for any reason, a Case Report Form (or a protocol for that matter) needs to be modified, to keep track of the version numbers. For example, the cover page of a CRF (or protocol) clearly highlights the current date and version number.

10.4.2 CRF Standardization

We have discussed trying to modularize the CRFs as much as possible. In general, one way to do this is to standardize the format and the data collection as much as possible. Thus, pages pertaining to medical history — inclusion criteria, exclusion criteria, and medication history — should have a format that doesn't change from protocol to protocol. As a reminder, use check boxes wherever possible.

The medication log pages for concomitant medication usage during the study require some comment. They need to accommodate multiple medications, given in varying schedules and via different routes of administration. Particularly vexing are insulin doses that might be varied on a daily basis or might be controlled by a pump.

Other forms used multiple times in an individual study's CRF include laboratory pages (although more and more studies are using central labs with computer reporting, so reports can simply be attached to the appropriate CRF page), adverse event pages, interim history pages, repeat physical exam pages, etc.

There also should be a completion / termination log page which makes it very clear why the patient ended participation in a study. This page correlates directly with the Patient Disposition table in the CSR. Here is recorded whether the patient completed all study procedures successfully or didn't. If the patient terminated early, the reason should be specified here. Usually a series of check boxes frames some potential responses. For some though, more explanation is needed. For example, a patient may have withdrawn participation. It would be good to know the reason: Transportation issues? Adverse event (should be a corresponding AE / SAE page completed)? Etc. The termination page should enable one to clearly understand the disposition of the patient during the study.

Although we emphasize modularizing and standardizing CRFs as much as possible from an individual study across the entire development program, there will be studies in which very specific items are being studied. Those items might not be evaluated in other protocols or could be assessed in other ways. In those cases, I would suggest the protocol and the appropriate study-specific CRF pages be developed concurrently to ensure adequate and accurate capture of the data.

To reiterate a previous point, I think Case Report Forms should be designed to capture only the data necessary for analysis. Do not collect information that might possibly, probably, sort of be used at some point, conceivably in the future.

10.5 Statistics

Now, a (very) few words about statistics. Keep in mind, I am not a statistician and, no matter what I do, I will not make the reader a statistician (assuming the reader isn't one already).

This section will simply point out some of the issues important in considering statistics within the context of performing a clinical trial and conducting a successful development program.

Let's start with a few fundamental premises: We need high quality data. We need data that are convincing. We need analyses which, at the end of the analytical process, are clear-cut in relationship to efficacy and to safety. Keep in mind that statistics can change our clinical trial design strategies, depending on what particular statistical approach we use.

In any successful development program, whether considering a single page in a CRF or discussing which statistical tests are utilized to analyze all data from all studies, the key is to ask the right questions. If the right questions aren't asked, the right answer will not magically appear.

10.5.1 Populations and Probability

The concept of probability facilitates our understanding of what is actually looked at or studied in a clinical trial protocol and contrasts that with "reality". Except in the case of very rare diseases, the population sampled in a clinical trial — regardless of the number of patients studied — represents a microcosm of the larger "real" population.

For example, the study might evaluate 20 patients, or perhaps 200 or maybe even 2,000. But each of those numbers represents a very small subset of the total patient population. The variability in those two, 200, or 2,000 patients may be somewhat limited compared to the variability in the entire population.

The assessments made in the study are based on the subset of the total population actually studied. Then, the information collected and analyzed is extrapolated to the larger population. At that point, the important question is how reflective of the whole the study subpopulation is; *it is not necessarily reflective of the whole.* There are additional elements of probability that are related to randomness and to bias, to certain environmental factors, individual factors, all of which will affect the ultimate answer that obtained in the study.

Certain statistical approaches and statistical tests are employed in an attempt to try to eliminate, as much as possible, those elements of randomness and bias. At the end, can we completely eliminate them? Can we be perfectly convinced that we have perfect correlation between the sample that we evaluate and the population to which we extrapolate? Most certainly not. But we need to reassure ourselves, to the maximum degree possible, that we have done the best job possible and that the data, within the limits of the methods we employ, are in fact extrapolatable to the larger population.

10.5.2 Metrics and Scales

The scales we use also affect our ability to extrapolate the results.

As we look at the scales, as we look at probability in a large sense, there are multiple scales that we can employ. We can use "yes / no" types of questions. We can use gradations: Is it mild? Moderate? Moderately severe? Severe? Etc. Some gradations can be particularly nettlesome, such as various coma scales. Or, we could use fairly straightforward designations: Either the person is alive or not alive. Scales used will affect the ease or difficulty with which the data can be extrapolated to the larger population.

10.5.3 Hypotheses and Errors

The approach to clinical trials starts with the null hypothesis. The **null hypothesis** (designated Ho) states that the starting premise — whatever the premise is — is true, unless the experimental data that we collect compel us to reject the particular hypothesis. The starting null hypothesis is that there is no difference between the respective study groups.

An **alternative hypothesis** (designated H1) might also be formulated — essentially the converse of the null hypothesis. However, the statistical approach is to either reject or fail to reject the null hypothesis.

Statisticians will also state that the null hypothesis cannot be either accepted or rejected unless all the elements that go into the acceptance and rejection are identified and defined at the beginning or the experiment or clinical study.

Since we are not sampling the entire population, we can make errors in our sampling, and therefore, our analytical processes. (See Figure 10-5.) Essentially, there are two types of errors we are subject to: There is type I error. In that case, we believe (based on the data we collected and analyzed) there is a treatment effect (i.e., the null hypothesis is rejected), but in reality there is no difference.

The alternative is type II error. In type II error, there really is a treatment effect, but we believe there is no treatment effect (based on the data collected and analyzed). Our analysis leads us to accept (or fail to reject) the null hypothesis, but we are making a mistake.

Figure 10-5. Error Types

	Believe there IS a treatment effect	*Believe* there is NO treatment effect
Really IS a treatment effect	OK	Type II error
Really is NO treatment effect	Type I error	OK

Type I and Type II errors are concerns in all clinical trials. We attempt to avoid the errors by increasing the sample size (to increase the power) and to increase the separation between treatments required to "demonstrate" statistical significance. Yet, we always balance those goals by the practicality of being able to complete the trial in our lifetime.

10.5.4 Statistical Principles

We have previously mentioned ICH guideline E9. It is a required read! Below I will highlight a few considerations that I believe are especially important in planning a clinical trial or a development program.

In preparation for trial design, as emphasized many times previously, consider the trial's place within the context of the total development plan. Define the nature of the trial (exploratory or confirmatory), the scope of the trial (considering population and variables), and the methods to be employed to reduce bias (such as blinding and randomization).

Other elements in the trial design to be considered include the configuration of the trial (e.g., parallel, cross-over, or factorial design), numbers of centers (single vs. multiple; single country vs. multinational), comparisons studies (superiority, equivalence, dose-response, etc.), the treatment sequencing, sample sizes (including subgroup analyses), and the methods and processes of data capture and processing.

Next, it is important to establish aspects of trial conduct such as monitoring processes, inclusion and exclusion criteria (and changes occurring during the study), the rates of accrual, whether or not an interim analysis will be performed (and what decisions will be made as a result of conducting the interim analysis), and the role of data monitoring committees.

Data analysis considerations also include defining the analysis sets (i.e., the analysis populations), determining the procedures for dealing with missing values, outliers, data transformations, trial size estimates (and assumptions on which the estimates are based), data interactions, data integrity and software validation, and the implications of these decisions for the analyses of safety and efficacy.

This chapter has been about data and data flow — starting at the end (in this case, the Clinical Study Report) and working our way backward through the integrated analyses, through approaches to individual studies, to definitions of the components of the statistical analysis plan, to the analysis of the data, and, finally, to the collection of the data on the case report form.

Chapter 11
Trial Safety

This chapter will focus primarily on safety issues associated with conducting clinical trials. We will look at adverse events, first from a regulatory framework and then from a monitoring perspective. Then, we will touch on safety issues occurring during the conduct of a trial or a development program.

Occasionally, trials don't work out the way they were planned.

As an example, during a Phase 1 study of a nucleoside analog for the treatment of hepatitis B, five patients died and several others experienced severe liver and pancreatic toxicity at several different clinical trials sites, in a study with a relatively small number of patients.

The FDA's response to this particular event was to issue seven warning letters, including to the investigators and the sponsor. The issues cited were: failure to report the adverse events, failure to obtain proper consent, failure of control of the trial by the sponsor, failure of control by the investigator and the institution, and failure of the IRB to adequately monitor the trial. As a result, new regulations were promulgated to deal with these potential problems in the future.

There have obviously been other disasters in the course of drug development. Some examples are: Zomax, Phenformin, Vioxx, and Selacryn.

Selacryn was a drug marketed by SmithKline and French. Selacryn had been developed as a diuretic and was very effective. During the course of the clinical trials, the drug appeared to be extremely safe . . . *except* . . .

I say "except," because there were a few instances of liver toxicity associated with the use of the drug. However, the reports regarding those events appear to have never made it to the database and, thus, were not contained in the data submitted to the FDA for marketing approval.

Fast forward, and following approval, the drug is being sold around the world. Clinically, it was very well received as a diuretic, but more reports surfaced about potential liver toxicity. As part of the review, it became apparent that SmithKline should have been aware of the potential problem. Following what apparently was an extensive investigation by the FDA and SmithKline, the drug was pulled from the market and a couple of people were very, very close to going to jail as a result of this particular event.

On the surface, it would appear that there was evidence of negligence on the part of SmithKline. Certainly, that is the way one would logically interpret the penalties: debarment of a couple of individuals and a corporate consent decree. Consent decrees with the FDA are generally ugly things. In the process of responding to a consent decree, the company needs to significantly improve its monitoring systems and change its corporate attitude regarding the types of events that got them in trouble in the first place.

In fact, at SmithKline, the consent decree resulting from this particular event led to a dramatic change in the psychology related to drug development. By the time I arrived, a few years after the events, the psychological and operational effects still pervaded the entire organization — not necessarily in a good way. For example, one of SmithKline's solutions was to implement a data management system that would have so many edits, checks and cross checks, that nothing like that event could happen again. The net effect was to essentially paralyze the data reporting system — simple reports would take literally days to produce. As a result, when working on my first NDA, the team ultimately had to devise alternative systems for production of reports.

SmithKline's approach was to have computers "solve" the problems and prevent future problems. Keeping in mind that the problematic data never actually got into the database, it isn't clear that was the right solution. However, the point of discussing this example is to illustrate a real-world situation and to indicate that in the process of drug development, there will be problems with drugs — *there is no such thing as a perfect drug.* Accordingly, judgment and integrity will always need to be employed.

11.1 Regulatory Framework

Trial safety, and in particular adverse events, should be viewed within a framework of regulations dealing with safety. The first reference document for this is the ICH E6 Good Clinical Practice Guideline (www.emea.europa.eu/pdfs/human/ich/013595en.pdf). Other applicable guidelines include the ICH Guidances E2A, Clinical Safety Data Management (www.emea.europa.eu/pdfs/human/ich/037795en.pdf) and E2B(R3), Data Elements for Transmission of Individual Case Safety Reports (www.emea.europa.eu/pdfs/human/ich/16678305en.pdf).

In the E6 Guideline, an **adverse event** is defined as "any untoward medical occurrence in a patient or clinical investigation subject who is administered a pharmaceutical product and this event does not necessarily have a causal relationship with this particular treatment". The Guideline continues, in the explanation of adverse events . . . "An adverse event can, therefore, be any unfavorable and unintended sign" — and the guidelines specifically include any abnormal laboratory findings — "symptoms or disease temporally associated with the use of a medicinal or investigational product whether or not related to the medicinal investigational product."

In the United States, 21 CFR 312 defines a **serious adverse event** as "death, a life-threatening event, inpatient hospitalization or a prolongation of an existing or current hospitalization, or any persistence of significant disability / incapacity or congenital anomaly or birth defect". The ICH definition is identical.

At a practical level one issue has to do with hospitalization. If the patient is currently an outpatient and the patient experiences an upper respiratory infection that is severe enough to result in hospitalization, this event would be considered a serious adverse event — even if investigator and sponsor viewed the event as being unrelated to the study drug. The hospitalization alone throws it into this category. Additionally, a similar categorization results when a hospitalization is extended by a day or two as a result of the occurrence of the event.

Life-threatening adverse events are defined in 21CFR 312.32. These are events that place "the subject, in the view of the investigator, at immediate risk of death from the reaction as it occurred in that particular patient. This definition does not include a reaction that might have caused death had it occurred in a more severe form."

Take an example of an allergic reaction. If the person had, for example, hives and the hives were pretty severe, but treatment with Benadryl® or epinephrine resolved them promptly, that event would not be labeled as a life-threatening adverse event. By contrast, however, if that same patient had hives and then had acute respiratory failure requiring respiratory support as a result of the anaphylactic reaction, that event would be classified as a life-threatening adverse event. Both events, though, would have represented an allergic response.

The next important definition is that of an **unexpected adverse event**. We have touched on this definition a couple of times previously. Fundamentally, it is an adverse event that is not included in the Investigator's Brochure.

There are several components to the definition that are worth considering though. Of course, if the event isn't mentioned at all, then it qualifies as being "unexpected". However, severity and specificity are also considerations. For example, we have noted previously, in either animal studies or in the initial clinical studies, that modest elevations of liver enzymes had been observed rarely at higher doses. Those observations are duly recorded in the Investigator's Brochure.

If a patient in a current trial had mild elevations in liver enzymes with no additional clinical sequelae, that event would logically be judged "expected". However, if five patients in a row were noted to have elevations in liver enzymes, the frequency of observation has changed; and the event qualifies as being "unexpected". Or, if a single patient has an elevation in liver enzymes, but the event requires hospitalization for hepatic necrosis, this severe event is not listed in the IB and is therefore "unexpected".

Disability is a term qualified by a person's ability to conduct normal activities. It reflects a person's normal activity level and capacity. (21 CFR 312.32) Given that each of us has different levels of "normal activity", a disruption in activity that would result in disability is necessarily dependent on the individual involved.

The final term to define is an **associated adverse event**. An adverse event is deemed to be associated if there is a "reasonable possibility that the adverse event may have been caused by the product". (21 CFR 312.32)

In real life, we are infrequently completely certain that an adverse event is related to the use of the product. On the other hand, if there is a degree of reasonable certainty, then we need to apply a label of "associated" to this adverse event. Different companies will deal with this differently on their adverse event report form. Different terms are used to deal with the uncertainty. For example, the gradations of association on the CRF might be "definitely", "probably", "possibly", "unlikely" or "not". With that set of gradations, investigators can feel more comfortable hedging their bets and dealing with what is likely to be less than complete certainty. The difficulty comes in determining what qualifies as being reportable to a regulatory agency. Most companies would put "definitely", "probably", and "possibly" in that category. Some, more conservative companies will even add "unlikely" to the category of "associated" and therefore reportable. The essential point is that if there is a reasonable possibility the adverse event may be linked, then it needs to be identified as being associated.

11.2 Trial Safety Monitoring and Reporting

In the United States, once an IND is open, the sponsor has certain responsibilities regarding updates and maintenance.

Information regarding adverse events and adverse experience must be provided to the Agency regardless of whether it derives from clinical trials, from non-clinical studies or from marketing experience in other countries. For example, any non-clinical animal studies which suggest significant risk for humans, typically toxicology studies, need to be submitted no later than 15 calendar days after the sponsor receives the information.

11.2.1 Annual Reports (21 CFR 312.33)

At least on an annual basis, an update must be filed to the IND summarizing all safety information (and appropriate efficacy information) obtained by the sponsor over the prior year. This annual summary includes any 7- and 15-day reports filed during the year and all other non-serious, expected, and non-related AEs. This information is largely summarized in tabular formats, but includes at least an interpretive summary reassessing the risk:benefit ratio of the compound in light of the new information.

Safety reports are also submitted at the time of serious events.

11.2.2 Expedited Reporting

There are three key components in the determination of expedited reporting: seriousness, expectedness and causality.

The assessment of **seriousness** requires an understanding of the difference between severe events and serious events. **Severity** describes the intensity of the event. For example, the event itself may be relatively minor, but severe as far as the patient is concerned; for example, a severe headache. **Seriousness**, by contrast, is determined by outcomes or potential outcomes. If an event results in death, is life threatening, requires hospitalization, results in disability or a congenital abnormality or represents a "medically significant" event, it qualifies as being serious. For example, if a patient presented to an emergency room with a headache and was found to have experienced a stroke, the stroke (although not the headache) would be a serious event.

Expectedness, as noted above in Section 11.1, refers to prior experience with the compound or related compounds and is largely determined by the Investigator's Brochure associated with the product and the trial or, for marketed products, the product label.. If the trial we are reviewing is the first clinical experience with the drug, we could argue that everything is unexpected — hasn't been seen before. An exception to this logic concerns structurally-related compounds with prior clinical experience. In this case, the IB might list reactions (including frequency and severity) with the related compound and argue that similar events with our compound would, in that case, be expected. For the first trial or two, the approach is useful, but with accumulating clinical experience we are obligated to specifically identify the experience with our compound. Then, the information in the IB related to experiences with our compound clearly constitutes the frame of reference for the "expectedness" determination.

Causality is the assessment of relatedness between the study drug and the event. See the discussion of Associated Events in Section 11.1. It is important to note that even if the investigator's assessment is disputed by the sponsor, the investigator's assessment is always listed in tabulations. There can be annotations about disagreement, but the investigator's assessment is *never* overruled or replaced.

Expedited reporting means reporting in a short-term time frame (e.g., 7 days or 15 days), rather than at the time of the annual report. (The discussion below focuses on reporting requirements in the US. It is important to note that reporting requires differ from country to country; thus, it is important to understand the local regulations when conducting trials outside the US.)

Essentially, it works this way.

It is *not expedited* (i.e., it goes in the annual report) if

- it is *not* serious
- it is *not* related
- it *is* expected

It *is expedited* if

- it *is* serious
- it *is* related
- it is *not* expected

The format for expedited reporting in the US is a MedWatch form. The form 3500A is now online and can be used as a worksheet and as a means of updating the report as new information becomes available. (http://www.fda.gov/downloads/Safety/MedWatch/HowToReport/DownloadForms/UCM082728.pdf)

A **15-day report** is required each time the sponsor receives a report of a serious and unexpected adverse event. This also applies to events that, while previously observed, are now noted to occur with increased frequency. Once such events are noted in the IB, they are no longer unexpected and thus would be accounted for in future annual reports.

By contrast, a **7-day report** is communicated via telephone (and subsequently in writing) for adverse events that are serious, unexpected AND fatal or life-threatening and are deemed to be related to the use of the drug (being mindful of the vagaries of this assignation — if unsure, report!).

The clock for both the 7- and 15-day reports starts at the moment ANYONE in the sponsor's organization (including CROs) becomes aware of the event.

While it is useful to have as much information as possible at the time of reporting, the 7- and 15-day limits are fixed. If necessary, follow-up information is provided. (By the way, in my experience, additional information almost always becomes available after the initial regulatory report and, thus, follow-up is required.)

In case there might have been a question, death is always serious (probably particularly to the individual involved). Even if previously identified as a potential consequence of participating in a trial with this drug product (for example, the drug is used for cancer, AIDS, or late-stage Alzheimer's), a 15-day report is required.

On the other hand, if someone dies in a study and there is something incongruent about that person's death compared to other deaths noted, it might make sense to expedite the report. A further consideration is looking at the frequency of events. If the rate of deaths seems to have escalated, even in an oncology study, perhaps the change would need to be reported in an expedited fashion.

These latter circumstances require judgment and having an adequate safety monitoring system in place. I will put in a pitch for having a safety review committee internally to review this information (if your company is small, this might include some external consultants). This is different than a data review monitoring committee, which is always an external group. Larger companies tend to have a group of people whose sole job is to monitor and review safety information in an on-going fashion.

It might seem obvious, but the report form must include adequate information to identify the patient (not by name, but by code — e.g., study 101, patient ID 106-103), the reporter, the product (drug [and dose] or placebo or blinded product [if blind not broken]), and the event.

11.2.3 Periodic Safety Reports

Periodic reporting must be made for products prior to approval (to the IND in the US) and post-approval (to the NDA in the US). The minimum requirement for IND reporting is annually. Reporting to the NDA is quarterly for the first three quarters after marketing and then annually thereafter.

Periodic reporting usually provides a tabular summary of the events as well as narrative summaries. There is always an analysis that puts the events in the context of all information available at the time of the report (e.g., changes in disease management, non-clinical studies, toxicities associated with other therapies, etc.). In addition, it is important to describe the sponsor's response to the events — changes in IB (prior to approval) or changes in product labeling (post approval).

11.2.4 Adverse Events — Trial Processes

In assessing and reporting Adverse Events in an individual clinical trial and throughout the development program, several documents and processes come into play.

11.2.4.1 Clinical Trial Protocol

For an individual clinical trial, the protocol is the first guide for all participating in the study regarding the processes for adverse event reporting.

The protocol identifies what must be identified and reported as AEs, identifies the classification process for the event, identifies the process for reporting the event to the sponsor and IRB (as appropriate), and defines the period of follow-up for adverse events that occur during the study.

In addition, the protocol should identify what should NOT be collected on the CRFs as AEs. For example, if a patient were noted to have a headache at the time of entry into the protocol, prior to dosing, and returns for a visit one week after initiating dosing and still has a headache, but that headache is no different than it was at the beginning of the study, it is not an AE and should not be recorded as such. On the other hand, if the headache were reported as mild at the time of study entry and now, at the 1-week visit, the headache is reported by the patient to be severe, the change in severity constitutes an adverse event and would now be recorded as such on the CRF.

11.2.4.2 Investigator's Brochure

As we noted previously, the Investigator's Brochure (IB) is the bible when it comes to the list of adverse events. If the AE is in there, the AE qualifies as being "expected". See discussions above in Sections 11.1 and 11.2.2.

At a minimum, the IB needs to be updated annually. However, it should be updated whenever "significant" information becomes available. The IB is an important reference for the clinical trial. It is not a document the investigator receives and then places on a shelf. It is a crucial tool in enabling the investigator to fully understand the drug product he or she is using in the trial and, in particular, the known adverse event profile of the drug product. (See Chapter 5 for a more detailed discussion of the IB.)

11.2.4.3 Case Report Form

The Case Report Forms represent an important element in the chain of information collection and reporting. They must be designed for ease of use and for adequate collection of information. In general, most studies will include case report form pages for both adverse events and serious adverse events. Occasionally, a third form for deaths is also included.

The CRF is a tool that must be understood by everyone involved in the trial: investigators, study coordinators, study monitors, data managers, regulatory personnel, statisticians, and regulators. Accordingly, all CRFs benefit from clarity and intuitive utility.

We have discussed previously about poor CRF design. Some of the common problems related to CRF design include: duplication of records or fields, page design and flow that is confusing, forms that don't include input modules for crucial information, too many text fields, and fields that provide output inconsistent with the expected output for statistical analysis.

11.2.4.4 Trial Database(s)

The next stage in consideration of the handling of AEs in the development of the database is to collect, organize, track, and enable analysis of the information related to adverse events. Expedited reporting of AEs generally precedes the collection and monitoring of the CRF pages. Thus, the sponsor must have a mechanism for tracking and cross-checking the events against information collected in the CRFs. All information must be reconciled prior to database lock.

Keep in mind that it is the sponsor's obligation to obtain *all* follow-up information on serious AEs until the events have either resolved or are stable. Accordingly, there must be close cooperation and coordination between monitoring and data management.

This type of information is vital to the construction of the Integrated Summary of Safety and the overall analysis of the risk:benefit ratio. As such, the database must enable thorough, adequate, and complete collection and reporting of information related to individual AEs, to all AEs in a given study, and to all AEs across the development program.

11.2.4.5 Data Monitoring Committees

Another wrinkle in the AE handling process is the interaction with the Data Monitoring Committee. An updated guidance was issued by the FDA in 2006 regarding operation of DMCs. (www.fda.gov/downloads/RegulatoryInformation/Guidances/UCM127073.pdf) If a DMC is in place for the study, they always have a concern for safety, regardless of other activities in which they might be involved. They provide an independent, hopefully objective, review of safety information, putting it into the context of the larger information related to assessments of risk and benefit. Need for a DMC for a given trial is generally determined by the perceived risk of the study and the potential to provide additional scientific validity to a study. The DMC is always operationally independent of the sponsor, although their expenses are covered by the sponsor.

11.2.4.6 Literature Searches

It is the sponsor's obligation to keep up to date with information about the compound, related compounds on the market, or compounds with related chemistry or mechanisms of action that are in development. Literature searches regarding this information should be periodically performed and included in IB updates (as appropriate) and in IND annual reports (as appropriate).

The net for capturing information potentially includes all reported non-clinical studies and all reported clinical studies whether in abstract or full report form. The scope is worldwide and for reports in languages other than English, translations must be provided. Obviously, this is more of an issue for later-stage compounds or for products already marketed.

See Table 11-1 for a listing of some United States Food and Drug Administration and ICH resources regarding adverse event reporting requirements and processes.

Table 11-1. Adverse Event Reporting Resources

United States		
	Adverse Event Reporting System	http://www.fda.gov/Drugs/GuidanceComplianceRegulatoryInformation/Surveillance/AdverseDrugEffects/default.htm
	MedWatch	http://www.fda.gov/safety/MedWatch/default.htm
	Enforcement of Post-Marketing Reporting	http://www.fda.gov/Drugs/GuidanceComplianceRegulatoryInformation/Surveillance/ucm129115.htm
ICH		
	Clinical Safety Data Management: Definitions and Standards for Expedited Reporting	ICH E2A
	Data Elements for Transmission of Individual Case Safety Reports	ICH E2B
	Clinical Safety Data Management: Periodic Safety Update Reports for Marketed Drugs	ICH E2C

11.3 Clinical Holds, Maintenance, and Reporting

11.3.1 Clinical Holds

A **clinical hold** results when the FDA orders a delay in the start of an investigation or orders the suspension of current investigations.

As you remember, INDs are never formally approved. Following the 30-day review period, if the sponsor hasn't heard from the FDA, the initial trial may proceed. Typically, however, the FDA will take its full 30 days to review the information and then contact the sponsor with several questions. Most of the time, those questions are procedural or details or they're informational and, as a result, they don't really hold anything up. As long as the sponsor's response is timely and addresses the questions, the trial proceeds.

On the other hand, the reviewing division may contact the sponsor and state that the trial may not proceed and the program is on clinical hold. Because of the formality of this declaration and the additional procedural activity that is required, the division will not do this lightly.

If the communication relates to the initial study proposed in the IND, no subjects may be given study drug. If the communication relates to an on-going study, no new patients may be given study drug. Whether those currently receiving drug must stop is dependent upon the situation and will be specifically communicated by the Agency.

In Phase 1 studies, the grounds for hold would be either an unreasonable or a very clear, significant risk of illness or injury that is not appropriately dealt with in terms of the protocol, monitoring, safety monitoring, follow-up and so forth. The Agency, as I also mentioned, could review the protocol and the program, look at the investigators and believe that their backgrounds are insufficient to conduct a particular clinical trial that's in place.

The Investigator's Brochure could, in the Agency's opinion, contain misleading, erroneous, or incomplete statements. The Agency could also conclude the IND just doesn't have enough information with which to make an assessment of risk:benefit for this first study. Manufacturing issues can also derail the IND and result in a clinical hold.

Phase 2 and Phase 3 studies are also subject to clinical hold if, in the opinion of the Agency, the on-going studies pose potential harm to patients. Any of the grounds mentioned regarding Phase 1 studies applies here as well.

New toxicology information, information provided to the IND regarding safety issues, or new clinical trial protocols submitted to the IND may also trigger a clinical hold.

New (additional) protocols deserve a bit more comment. Even though the sponsor does not need to wait to initiate additional clinical trial protocols (after the IND has gone into effect), the reviewing division will look at all materials submitted to the IND, including new protocols. If the division thinks the design of the protocol is insufficient to accomplish the goal in conducting the study or that the monitoring process of patients, particularly in terms of safety, is insufficient to be able to identify potential problems associated with the use of the product, the study might be put on hold.

Other items can provide grounds for clinical holds. Study design might seriously impede enrollment, such that there is little potential benefit to the few patients actually enrolled. There may be insufficient quantity of drug available to complete the study, or reports may have been filed with the Agency regarding previously conducted studies suggesting there is no benefit to the drug.

Another drug in the same indication might demonstrate a much superior risk-to-benefit profile. Although theoretically possible, it is highly unlikely that the Agency would actually impose a hold for this reason.

If the Agency believes the sponsor is not actively pursuing the marketing application, a clinical hold may result. Given a lack of minimal activity, this type of decision shouldn't be a surprise to the sponsor should it occur. On the other hand, if other reasons have interfered but the sponsor is really planning to initiate trials, it behooves the company to maintain contact with the Agency to ensure the clinical hold doesn't happen.

The FDA may also well determine that the study is not in the public interest. Previously, we have some of the potential issues related to inclusion of women of child-bearing age in clinical trials. Pediatric studies also represent an important discussion point in clinical trial programs. As an example of the potential impact, it is possible the Agency may stop or slow down a program because we've written the inclusions / exclusions criteria in such a way that we've arbitrarily essentially excluded subjects of child-bearing potential without an adequate explanation.

11.3.2 IND Maintenance

Several activities must occur on a regular basis to maintain an IND.

Any time protocol changes are enacted, the changes (in the form of a protocol amendment) must be provided to the Agency. New editions of the Investigator's Brochure and information regarding added or dropped investigators (on Form 1572s) need to be provided to the IND as available.

As noted above, if putting a new protocol in place (after the initial protocol submitted as part of the IND has been reviewed), that protocol must be submitted to the FDA and be approved by the appropriate IRBs. (Once IRB approval is obtained, it is not necessary to wait for FDA approval prior to initiation.)

It is common (particularly for protocols with extended observation periods) that modifications to the protocol will occur during the study. Sometimes, the changes are simple and straightforward — clarifying wording. Those administrative changes can be filed to the IND "in due course".

Other changes might be more substantive; for example, changes in the inclusion / exclusion criteria, increases in frequency of monitoring, changes in dosage or dosing schedules, or alterations in the randomization schedule for safety reasons. These changes should be approved by the IRB prior to adding more patients into the study and thus prior to implementation. In the filing to the IND, the cover letter should clearly identify the reasons for the changes and that IRB approval was obtained prior to implementation. Depending on the significance of the changes and the reasons for implementation, the cover letter might also address the risk:benefit ratio in view of the changes and why the trial should continue with the changes in place.

If a new investigator is added to the study — either at a new site or at an already functional site, the FDA needs to be notified within 30 days of adding the investigator. In this case, that 30-day time period is typically the point in time when the sponsor provides the site with study supplies. The spirit of this rule is "timely notification" of the FDA regarding investigators. If the sponsor is conducting a large, multi-center study, the investigators will typically come on line at staggered time points. The FDA doesn't like to get new amendments every other day, so it is reasonable to group the new investigators together and file every 30 days or so until all investigators are enrolled. The Form 1572 is completed with the protocol number, investigator's name and his or her qualifications. If there are multiple studies being conducted with the investigative product, ensure the forms are clear in differentiating the studies.

Other information must also be submitted periodically as part of the IND maintenance process. For example, information amendments contain additional information, usually reports, that need to be provided to the IND, but aren't specifically safety issues or tied to new protocols.

Such items might include reports dealing with toxicology studies, CMC studies or process changes, stability reports, etc. As a procedural matter, these items are generally bunched in order to avoid a lot of little submissions. They also, depending on the appropriateness of the timing, might be included in the IND annual update submission. The key item regarding the timing of submission is that the events DO NOT negatively impact the safety profile of the product.

11.3.3 IND Safety Reports

Written reports are required within 15 days — as we noted earlier — for AEs that are serious and unexpected, and for non-clinical animal studies that modify the human risk profile.

Email, telephone and fax reports are appropriate for unexpected events that are fatal or life-threatening and — again as we noted earlier — must be filed within seven (7) days. (Although emails and fax reports satisfy the reporting obligation, it is, in my opinion, important to speak with either the project manager or the medical reviewer at the division responsible for your IND.)

It is important to note, in all cases regarding expedited reporting, that the initial communication is just that — an initial communication. The expectation is that not all information will be known at the time of submission (although one likes to provide as much as possible) and that additional information will be provided until the event has resolved (i.e., death, resolution, or stability of the event).

Although extensive discussion of IND Annual Report formatting and submission is beyond the scope of this book, we will mention a few items that of crucial in the context of clinical trials. The regulatory framework for the Annual Report is contained in 21 CFR 312.33. The anniversary date for the IND annual report is the date when the IND went into effect. For example, the IND was submitted to the Agency on June 1. No comments were received from the Agency, so the first patient was enrolled in the study on July 1. July 1 is the date the IND went into effect and, thus, becomes the anniversary date.

There is some flexibility around the submission date — 60 days on either side for the Annual Report. If the sponsor has more than one IND open, try to arrange to have them updated at the same time to reduce the internal workload. If the dates are a bit more apart than the 120 day window allows (60 days before and 60 days after the anniversary date), discuss it with the Agency — even if INDs are in different divisions. The Agency is generally amenable to coordinating the filings.

Individual study information is included. If there are multiple studies for which reports are being provided, I would suggest putting as much in tables as makes sense. It cuts down on paper (or electrons) and generally is easier to understand.

Information required includes the study title, the purpose of the study, the patient population being studied, and the current status (i.e., open / closed; if open, how many patients enrolled, completed, still in protocol, drop-outs, expected completion date, etc.). Again, much of the detailed information regarding patient status is included in tables for ease of representation. If the study is completed but not yet formally reported (hence, the final clinical study report cannot be included), then a brief but meaningful summary should be provided.

Annual reports will include updated tabulations of AEs (including most frequent and serious events), summaries of Safety Reports submitted in the prior year, lists of deaths (including cause and a summary), lists of drop-outs due to AEs (regardless of association with study treatment), lists of non-clinical studies completed and in progress (and known results), and any new, relevant information regarding activity of the study product.

Also, important CMC changes must be included and described. Their impact (or lack of impact) on the clinical trial material must also be discussed. In addition, newly available toxicology information needs to be included and assessed.

Annual reports require inclusion of the general investigational plan for the upcoming year. The plan tends to be more specific for the immediate future and less specific the further out the time frame being considered.

This is a good time to revise the IB if it hasn't been done very recently. Obviously, if the IB were revised one month before the Annual Report is being submitted, just include that one. But, if the most recent version is nine months old, I would suggest updating it.

The Annual Report might also be a good time to review progress on the Target Product Profile that had originally been agreed to with the Agency. If there are new questions generated by new data, this might be a good time and place to identify the issues and engage the Agency in a discussion.

11.4 Risk Management and Safety

Risk management is the identification and implementation of strategies to reduce risks to the individuals who might be participating in our studies and the patient populations who might be exposed to our drug.

Very simply, one begins by considering events that have already been noted in non-clinical studies and prior clinical studies and developing a strategy for dealing with them in the future. It is also important to intelligently speculate about possible future events and how they might be dealt with should they occur. This isn't really designed to be a cover-the-waterfront exercise, but rather one that identifies reasonable, even if potentially remote, possibilities of adverse events.

Once they have been identified, then they are prioritized in terms of likelihood and in terms of severity. Plans are then designed to deal with the potential events. The extent of the details is driven by the likelihood and the severity. Obviously, the most likely and the most severe events should have some significant thought applied to the process with which they are handled.

The actual risk management plan should identify the risks and their likelihood and severity in writing. Then the plans are likewise committed to writing. It is important to decide who is in charge and how decisions will be made and responses implemented.

Above, we suggested establishing a safety committee in the company. While it is likely to already be in place in the larger company, as I mentioned, I think the smaller company should also have a group identified whose purpose and function is to review safety data — across protocols and across the program as a whole — as they become available. Implementing the risk management plan is a good place for them to be involved. It is particularly important to have such a group for serious issues (e.g., deaths that were unexpected) or completely unexpected issues (think Johnson & Johnson and the tainted Tylenol® crisis in 1982).

Chapter 12
Resourcing

12.1 What do you need to get the job done?

In this chapter we will focus on resourcing for development tasks and will discuss, in particular, a process for outsourcing, particularly to contract research organizations.

Remember that all organizations, no matter how large or small, are resource constrained. As a result, prioritization is always required. In *all* circumstances, prioritization depends on the time required to get the product to market, the size of the market into which the product will be sold, the overall set of resources available to work on the project, and the other projects within the organization that will compete for resources.

Although large companies primarily have an allocation problem, i.e., there are a lot of resources, but there are a lot of projects, in smaller companies, there just might not be enough resources or at least the right kind of resources to accomplish the task at hand.

Regardless of the environment, decisions regarding resourcing strategies must always be active decisions and are generally best if made in the context of pre-planning.

In large and small companies, a decision regarding potential outsourcing of projects or portions of projects must be considered. It is important to keep "on the table" a full range of options. One might hire additional resources. An alternative is to hire some and rent some. The other extreme is to contract out all work. As a reminder, the latter option, while potentially viable for a specific organization, nevertheless requires management.

12.2 Contract Research Organizations (CROs)

Enter contract research organizations (CROs). How do you decide to outsource and, if you do, what is the process through which the outsourcing is accomplished? In the discussion below, we will make the assumption that at least some activity is outsourced.

Initially, let's consider the "virtual" or "nearly virtual" development scenario. This presents the opportunity for virtual or "nearly virtual" product development. Let's see what we might have to consider.

One set of beliefs might be verbalized this way: "If I had my choice, I would do everything myself and then I would know that it would be done correctly." That probably applies to most of us — at least some of the time.

On the other hand, that very wise person, Anonymous, once said, "Management (or Leadership) is the art of getting things done through others." "Maybe I can do it better, but perhaps a team can accomplish more." Obviously, these are two different approaches to resourcing.

What are some of the common objections to outsourcing? One is, "We know our product best, so we need to run everything." Another one is, "No one understands our strategy the way we do." Of course, that always assumes the sponsor actually has a strategy. A third one might be, "I don't want a consultant talking to the FDA (or any regulatory agency) instead of me." Or, this is a particularly good one: "I have my own way of developing a drug." or, "Drug development strategies are just too delicate to entrust to a consultant or some other organization."

I have, in my consulting career, heard all of these objections to outsourcing.

Regardless of which approach the sponsor takes, though, I believe a set of clearly-stated goals are needed for the project. How about these? "We want rapid approval. We want rapid and extensive market penetration. In the end, we want to have made a whole lot of money." Those sound pretty good to me!

Table 12-1. Approach to Outsourcing

Why?
Preparation
Selection
Operations
FDA Regulations
Review

Again, let's assume you are at least willing to consider outsourcing. How might we approach it? (See Table 12-1.)

12.2.1 Why Outsource?

Why does one outsource in the first place? I maintain that outsourcing is really ultimately a business decision. Most people look at it as a resource decision: "I don't have enough people." or "I don't have enough materials within my company to do this particular type of activity." However, I look at it as a business decision and I will explain that shortly.

One might ask a question, "Should I outsource my work?" Phrased this way, the question is a very self-oriented question.

A somewhat better question might be, "Would outsourcing my work help my company obtain more regulatory approvals and reach world markets faster?" Now, refer back to our goals: Market penetration, achieving regulatory approval, and ultimately, making a whole lot of money.

The name of the game — the drug development process or device development process, diagnostic development process, whichever is being played — is, ultimately, how long does it take to get to market?

What benefits might accrue to outsourcing? One possibility is that the combination of the Company and the outsourcing partner(s) might be able to shorten the time to market.

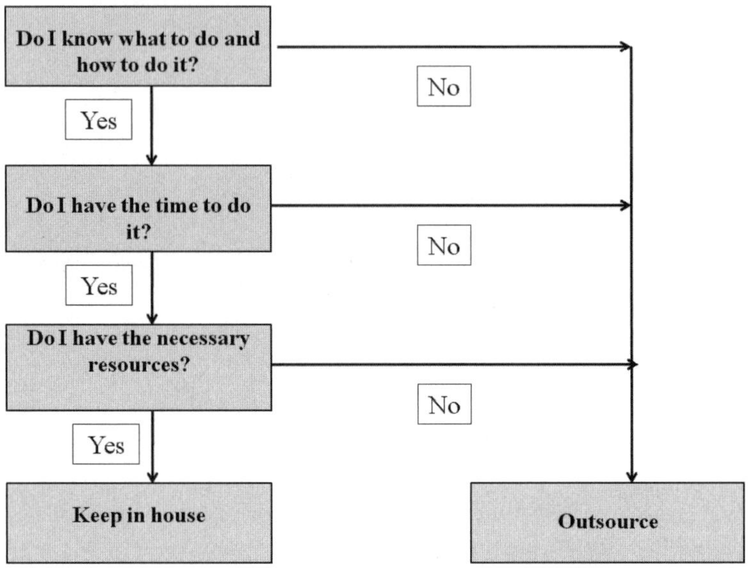

Figure 12-1. Outsourcing Decision Tree

How do you approach the decision? Figure 12-1 is one way of looking at the question, an outsourcing decision tree.

Starting at the top left, "Do I know what to do and how to do it?" If the answer to that is "Yes", then you move down to the next question.

"Do I have the time to do it?" If the answer to that one is "Yes", then I ask:

"Do I have the necessary resources?" If that is likewise answered "Yes", then we keep it in house.

On the other hand, if you answered "No" to any one of these particular questions, then you are driven toward the right side of this tree — that is, outsourcing the particular development process you are looking at. Keep in mind that it could be the entire process or only certain segments.

All of these decisions, however, need to balance three things: know-how, time, and resources.

"If I am going to consider outsourcing, what is the business value of that decision?" "To generate business value, is outsourcing a (the best) way to accomplish the goal?" These are complementary questions that drive the decision.

I argue that the sponsor is driven on the one hand by questions of capabilities, know-how, time and the ability to have the resources in hand for all the activities. On the other hand, it is well known that large pharmaceutical companies also contract work out to other companies. For the large companies, the strategic decision is the same — the best application of the best (or most appropriate) resources to accomplish their goals.

So, I suggest that the key framework is "**Business Value**".

Being mindful that the goal is to complete this project and get it to market (always remembering that "market" might mean partnering, etc.), why outsource?

Time, of course, is first; there is never enough time. Your company may have insufficient expertise, insufficient staff, or insufficient facilities. Obvious examples might be requirements for toxicology and other animal studies or manufacturing facilities; even larger companies are now outsourcing this type of work.

Another reason to outsource might be a strategic commitment to "virtuality". A company's strategy might be to avoid building much brick-and-mortar, therefore requiring that essentially all resources needed for the project will be purchased in the marketplace. The task, then, is managing the resources efficiently and effectively.

Financial considerations also figure into the "why outsource" question. For example, "When is it appropriate to replace fixed costs with variable costs?" Consider the hired worker versus the contractor. The permanently hired worker is paid a regular salary and receives additional benefits and represents a fixed cost. The contractor works a variable number of hours, requires no benefits, and can be terminated with minimal notice; this resource is a variable cost. The company can shed variable costs relatively easily; fixed costs are, in general, much more difficult and painful to delete. The argument also applies to required activity like manufacturing: Should the service be "rented" from a contract facility or should the company construct its own manufacturing facility?

Another key financial question relates to the speed with which we can get this product to the market. If we get a faster introduction, we get a quicker time to peak sales and, ultimately, we have a longer patent life, while at the same time having lower operational costs because of the shorter time frame. Consequently, we accumulate greater net aggregate revenues.

As we have noted previously, begin with the end in mind. So, if the decision is to outsource, planning must be done.

Understand exactly the reasons for outsourcing and quantify precisely what will be achieved, measured by time-based objectives and milestones. A performance scorecard can indicate those objectives that were met and those that weren't — with regard to time, resources, money, and milestones.

Next, we consider a set of questions that look a lot like the questions we asked at the beginning of the project.

- What is the therapeutic target?
- What is the best dosage range? Do we know the best dosage range based on currently-available information, or will it require additional studies?
- What is the regulatory environment at this point in time? What do we anticipate the regulatory environment will be at the point in time when we have completed the development?
- Is there a well-defined product business model for this particular product, for this particular class of product, or for the therapeutic target that we are trying to hit?
- What about the competition? What's the pricing that's already in the marketplace? What might be available at the point in time that we come into the marketplace?
- What, ultimately, do we believe the financial return will be for this particular project? *Then, will outsourcing change that equation?*

Ultimately, the question about financial return is the key to the outsourcing decision. Does outsourcing benefit us? What is the best business value?

These questions are as important in considering outsourcing as they are in developing the plan in the first place.

Above, we introduced the concept of virtuality. One version is what I call *Extreme outsourcing.* When I first started lecturing about this nearly 10 years ago, the idea seemed pretty radical. Now, it is not only not radical, it actually isn't uncommon. Extreme outsourcing is a paradigm in which a company really has minimal number of people within the company. They have relatively few owned resources other than an asset, cash and management. And their intent is to develop products, purchasing resources on the outside essentially for everything. Virtual is the name and outsourcing is the game that we are playing in this particular model.

In general, the model operates with a skeleton expert senior staff. Typically, it includes a CEO, a director of clinical, and a director of regulatory. Although I use the term "director", they could have very fancy titles. The point is, however, that the company has individuals with expertise in development — spanning regulatory, clinical, manufacturing, and chemistry. This core team will then make all of the decisions about outsourcing and will manage the outsourcing partner as well. Clearly, fixed costs are minimized in this model; there is no organization to build out. Dedicated facilities are kept to the minimum. The most extreme version of the model employs only management resources — all other resources are contracted.

In the *not-so-extreme model* — in fact, the most common model — a company picks and chooses which elements of the development operation to contract out. As I mentioned earlier, even the largest pharmaceutical companies in the world will contract out portions of development. Everybody will outsource something at some point in time during the course of the development, without necessarily committing to outsource the entire development process.

Sometimes it's appropriate for the sponsor to delay hiring specific groups within the company until there are more clear-cut signs the product will be successful. In that instance, individuals with very specific competencies can be engaged as consultants. They complement the core competencies already in the company. This "mixed" model enables great flexibility in modulating the company's burn rate (overall expense rate) and "buy" decisions can be delayed until more data are available regarding the product's likelihood of success.

By slowing down the burn rate — changing more costs to variable rather than fixed — the financial risk is decreased in case of project failure. I am aware of examples of companies who bet the other direction. In one case, the company built a manufacturing facility for their product; unfortunately, the product failed in Phase 3. Another example was a company that decided they wanted the best people (certainly not a bad goal). They achieved the objective, but they also fully staffed the entire organization; unfortunately, their product failed also. In each case, the companies invested a huge amount of money in fixed costs and the end results were ugly! In whatever way the organization is assembled, it is useful to maintain flexibility. Wise decisions with capital expenditures will also reassure investors.

12.2.2 Preparation for Outsourcing

How do you prepare to outsource? Figure 12-2 is a reminder of the pharmaceutical product development process as a whole. Any activity or process from the far left of this figure to the far right can be outsourced to some contract organization.

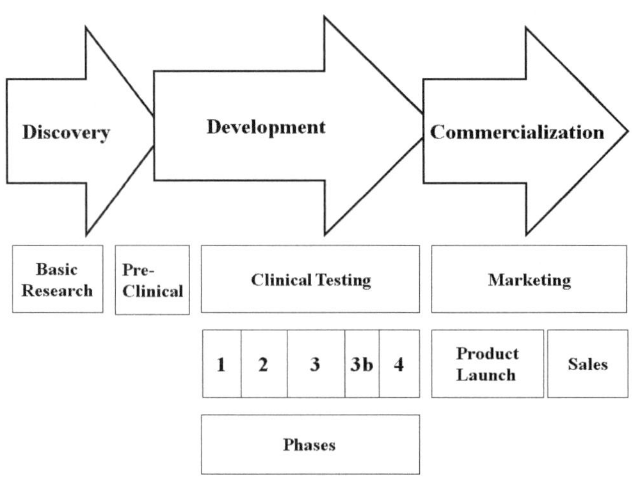

Figure 12-2. Pharmaceutical Development Stages

While each company is unique, all companies in the pharmaceutical business operate within a range of contexts — some specific to the company and some in the external environment. We discussed these in detail in Chapter 1. As for the development process in general, outsourcing also forces the company to have answers to certain questions prior to determining the specific outsourcing need(s). Such questions include knowing the company's business, the target markets, the competitive landscape, timing of projected market entry, and the current and future regulatory environment.

A crucial context is the financial environment for the company. For smaller companies, it is likely they will pursue paths other than product registration and sales / marketing of the product. Thus, timing requirements for funding become important considerations. What are the milestones that must be met to obtain new financing or partnering? Importantly, will outsourcing assist in that process?

We noted early on — in the discussion of plan development — the need to assess internal capabilities. Consideration of outsourcing gets to the heart of that assessment. An honest assessment of internal capabilities (including quantity and expertise) enables an honest decision regarding hiring or outsourcing or combining the two to optimize organizational and operational effectiveness.

Table 12-2 provides at least a basic set of capabilities and functional activities to assess as a first step in the process of considering outsourcing as a business value-enhancing proposition. Is the scope really a single study or the Phase 3 studies only? Is the scope the full program (toxicology, regulatory, clinical, data management, etc.) or only components of the program (e.g., clinical and data management only)?

If the program deals with development of a product for an orphan indication for which the total development program includes only 250–500 patients, the resources needed are obviously different in capability and number from a program dealing with rheumatoid arthritis that might ultimately need to include 3,000–5,000 patients in the regulatory submission package.

Table 12-2. Assessing Internal Capabilities (examples)

Market research (US, EU, Worldwide)
Regulatory and Quality Assurance resources
Medical and Monitoring resources
Data management resources
Project management experience
Development experience
Process validation experience

While the orphan program might be reasonably contracted out as a package — the same vendors will participate from beginning to end — the rheumatoid arthritis program might be considered in segments with decisions being made based on results obtained in the prior set of studies. While the full project needs to be managed in an integrated fashion, individual operational decisions might consider funding and outsourcing in smaller units.

The end product of the internal capabilities assessment within the context of the scope of the project is a set of written responses to the key questions: What are the objectives and timing of the project? What activities and capabilities are required to accomplish the project? What capabilities and resources are needed that are not currently available within the organization? Therefore, (assuming there are needs) what are the needs and when do I need to fill them?

12.2.3 Selection

Now, the easy part. How do you select a vendor? Okay, maybe it isn't that easy. A key to success is systematizing the evaluation and decision-making process. Start by generating a list of potential vendors.

Refer to the results of the internal capabilities assessment produced at the end of the last section. Plug the list of needs (activities and capabilities) into a list of large-category technical functions. See Table 12-3 for some examples.

Table 12-3. Outsourced Technical Functions

Discovery	Clinical studies
Formulation	Final product (scale up and manufacture)
Pre-Clinical	Regulatory submissions
Clinical Product (scale up and manufacture)	Marketing

Based on this information, identify the universe of relevant providers who might satisfy the capabilities you require. Input the list of needed capabilities (both large-category terms and specific capability and activity terms) into your favorite search engine and generate lists related to each activity or capability you searched. Also, using the same terms, consult lists of service providers developed by different organizations, e.g., the Drug Information Association (DIA) and the Regulatory Affairs Professional Society (RAPS).

Once the initial list is assembled (I would suggest that the initial list be as broad as possible), request background materials from the providers identified. Look at their materials, focusing on capabilities and experience. Once the information is reviewed, then short-list the potential contacts for possible meetings. At some point, check with friends and colleagues who might have had similar needs and particularly check out providers receiving good reviews.

Review the materials for *relevant* competence and experience. I would suggest that relevant experience is important. There certainly might be an occasion on which you are willing to take a chance on someone (individual or group) who is doing this for the first time. For example, they might have a lot of experience and understand all the components of the required activities, but just haven't done this particular piece before. Be open, but be cautious.

Once some candidates have been identified and that list has been narrowed to a group of interest for interviews, it is time to move to the actual selection process. Components of the process are listed in Table 12-4.

Table 12-4. CRO Selection Process

Prepare RFP (Request for Proposal)
Identify potential CRO candidates
Define and stick to a schedule
Submit request to CROs on candidate list
Review RFPs and select finalists
Audition finalists (in person)
Select

12.2.3.1 Preparing the RFP

First, put together a **request for proposal** (RFP). This is a formal written document that identifies in detail all of the services being contracted. The services need to be characterized as fully as possible. The most straightforward presentation of the functional activities being requested is in the form of a check list.

For example, the major category for assistance might be clinical trial conduct. Below that header is a list of all the activities related to preparing for, initiating, conducting, and completing a clinical trial. Those items are listed in the rows going down the left side of the page. In adjacent columns are checks (or Xs) to signify which activities are done by the sponsor, which by the contractor, and which by both. See more detailed discussion below.

Address a letter (or email) to the appropriate contact persons at the contract organization requesting a bid on the attached list of services. Identify the realistic schedule to which the RFP process is working and request that the RFP be returned by a certain date. Assuming the date is realistic, this provides the first inkling of the performance of that organization. By sticking to the indicated schedule, the sponsor will likewise convey the message that it is serious about time lines.

Once information is received, identify the individuals or organizations to meet face-to-face or to interview in detail (this might be conducted over the telephone). I would always recommend face-to-face meetings — it is good to look them in the eyes while discussing the project. Although time and hassle might be saved by meeting at the sponsor's location, there is also value in visiting the contract organizations at their location. The latter provides an insight into the dynamics within their organization. Sometimes by just simply walking around in a facility, one can get a very strong impression. The impression might be positive: "Whoa, look at these people. They are energized. They are enthusiastic. I'll bet they will do a good job." Or negative: "Whoa, these people are angry. They are obviously not going to do a good job for us." Either process (telephonic or face-to-face at either the sponsor's or vendor's site) can work, though.

Now let's drill down into the important components of the RFP process. See a proposed RFP organization in Table 12-5.

Table 12-5. RFP Organization.

Cover Letter
Time for questions and contact person for questions
Due date
Process review timetable
Outline protocol (better, attach protocol)
Definition of responsibilities
Prescribe format for response — especially the financial schedule (it makes it easier to compare bids)

The cover letter will define what it is that is being sought in the RFP. A key aspect here is the due date for this particular project. Several dates are provided to the prospective vendor. The first date is the date until which questions would be answered. The next date is the due date for the initial RFP. The next date is that for the first round selection. Finally, provide the range of dates for face-to-face meetings and the final selection date. In addition, it is important to provide putative project start and completion dates. Remember, the contractor's adherence to the timetable will provide the first hint about their performance ethic.

Attach to the RFP an outline of the project being outsourced — be as specific as possible. If it is a non-clinical or clinical study, the protocol should be provided. (It is best to provide the full protocol with the RFP. If it is necessary to start with an outline, realize that the bid might change significantly once the full protocol is provided.)

Define the responsibilities clearly. For example, you may be establishing a process for clinical development in which the sponsor will do the internal monitoring, but the CRO is to collect all the adverse event forms and review all the information prior to the time that it is returned to the sponsor. As the check list is compiled, identify the respective responsibilities as completely and specifically as possible.

Prescribe the format for response, particularly the financial schedule. Having been on both sides of the RFP process — as CRO providing the service and as company obtaining the service — I think this is an important point. If the prospective bids will involve quotes with detailed financial statements, I would suggest providing the template you would expect each potential vendor to use. This makes it much easier to compare the details of the bids as precisely as possible.

Table 12-6 is a simplified format of a check list of respective responsibilities.

Table 12-6. Check List of Responsibilities (Examples)

TASK	SPONSOR	CRO	COST	TIME LINES
SOPs to be used	X		NA	NA
Prepare Study Synopsis	X		NA	DD MMM YYYY
Write Protocol		X	S	DD MMM YYYY
Prepare Informed Consent		X	$	DD MMM YYYY
Arrange Insurance	X		NA	DD MMM YYYY
Etc.				

The tasks go down the left side. The next two columns specify the respective responsibilities. The latter two columns are optional — they can be included as guidelines, as hard-and-fast limits, or not included at this stage. It is always good to specify the general expected time lines, but you might choose to be less detailed in the initial RFP. Costs might not be specified in this detail at this stage. If not specified here, the detail is certainly required for the final agreement.

If the sponsor is requesting assistance with a clinical trial protocol, for example, the list of tasks to be specified might cover a broad range of activities. For example, site management might be specified and this might include site identification, site qualification, site visits, supply management, and document management. All of these items (including subtasks) need to be specified as clearly as possible. The process for adverse event handling might also be specified and responsibilities assigned. One sticky issue sometimes is management of contracts at the trial sites. Will the sponsor be handling that or will the CRO be handling all items, including negotiation (the Company, of course, retains signature authority)?

Other contract issues to be specified (as appropriate) in the RFP include describing who will be responsible for the clinical study agreement, the confidentiality agreement, and the informed consent template.

From a regulatory perspective, who is going to collect all the documents? Examples include the investigators' CVs, the protocol acknowledgment forms, the laboratory certificates, the laboratory normal tables, and so forth. Specify in detail the respective responsibilities for the elements defined to achieve project clarity.

Laboratory services is another area to consider outsourcing. Are you going to use local laboratories, meaning laboratories at the individual site? For example, you might use hospital A, hospital B, or hospital C. Each of them has a laboratory. If you use their respective laboratory facilities, you need to collect normal ranges for each one of those laboratories. Your data management is going to have to take into account the potential variations in the lists of normal values from laboratory to laboratory. That may work fine and that may be the appropriate thing for your particular study. However, you may want to consider use of a central laboratory so that you are dealing with a single set of normal values, a single set of processes, and so forth. Whichever laboratory arrangement is chosen needs to be clearly specified at the beginning of the RFP process.

Data management might also be part of the RFP. A data management plan, regardless of who is executing it, needs to be specified in the RFP. Who is going to do which elements of that particular plan? How are CRFs going to be tracked? Who is doing the data management per se? How will status reports be provided? Is the sponsor generating them? Is the CRO generating them? What detail(s) would be included? Who is doing the quality control? Who is doing the data import and transfer? Who is handling the database? How many different people or organizations are involved in putting data into that database? What is the method and format of laboratory data transfers? These are activities and decisions that need to be specified in the RFP as you consider issues of data management.

For project management, what services are required and who will be providing them?

This list of questions isn't meant to be exhaustive. It is meant to provide some insight into the RFP process and the detail needed to be specified to optimize the chances for project success.

12.2.3.2 Potential CRO Candidates

Next, consider the list of potential CRO candidates. Take a look at size and geographic reach. Your development program might be confined to the US, be primarily located outside the US, or possibly involve multiple regions. It would make sense that the final candidate has experience in the relevant geographic regions.

Match the needs, but never underestimate the requirements. Occasionally, once the project starts, the sponsor realizes that additional work should be contracted out. It is best to admit it as soon as possible and initiate a change (of scope) in order to extend the services requested.

Review and understand the CRO's relevant SOPs and their work flow processes. Also, review the technical aspects of whatever service is being contracted. For example, if data management is being contracted, an evaluation of the vendor's computer systems, security, validation, etc. should be conducted prior to signing a contract.

Table 12 - 7. Specify Trial-Related Duties in Writing

Task	Sponsor	CRO
SOPs to be used	X	X
Prepare Study Synopses	X	
Write Protocol		X
Prepare Informed Consent		X
Arrange Clinical Trial Insurance	X	
Etc.		

Table 12-7 describes how the SOP issue comes into play as one considers contracting out writing tasks associated with a trial. The SOPs used could be the sponsor's, the CRO's, or a combination of both. Just ensure that whatever processes are followed have documentation for them and are effective.

During vendor site visits, make sure there is an opportunity to meet and talk with the team the vendor is proposing for the project — certainly, at least, the key members. We all know the stories about the "bait and switch" routine of some particularly large CROs. You need to try as best you can to manage that process and to be able to retain the people who initially start on your team. It is possible to address some of this with contract language that imposes penalties for "significant" switches, but be careful — you also don't want to be saddled with someone who turns out to not be so good.

One important aspect — not easily quantified — is personality. This is a topic that shouldn't be overplayed or underplayed, but at least shouldn't be ignored. If, during the discussions between sponsor and vendor, there are conflicts between certain individuals on the respective teams, consider whether the conflicts are due to differing ideas or opinions or are more deep-seated personality conflicts. If the issues relate to personalities, it is very likely those issues will *not* go away (and will probably be amplified) during the course of the project. The *only* choices are to rearrange team members on the two sides (if possible) or to find another group with which to work.

I can't emphasize enough that you go through this process of preparing to work with the CRO. These projects are the lifeblood of the sponsor's organization. This might be the sponsor's only project. If so, this is the ballgame, so to speak. The project must be handled well. Make sure the people you engage are the right people and that they believe in the project. While the outcome of the project — success or failure of the drug in the trial, for example — can't be guaranteed, all effort must be expended to ensure that the project process is successful.

12.2.4 Outsourcing Operation

So far, we've talked about the preparation and the planning for the external contracting process. Now let's talk a little bit about how it actually operates in real life.

Appropriate budgeting is a key challenge. In smaller companies, in particular, there is a tendency to underbudget. It usually derives from two sources. The first is an incomplete breakdown of the tasks involved in the project. While almost everyone identifies the correct major topic items, the challenge is to go into enough detail with the tasks to fully understand the activities required. If you don't identify it, you are not going to budget for it. The second source is inadequate costing information. In this case the cost associated with a task is too low, either because the company didn't do enough homework to really understand the costing structure, or because they intentionally underestimated so the total cost of the project would seem lower than it actually is. There are few things that get people more upset in managing projects is having "out of control" costs. Many times, the problem was the initial planning.

Thus, it is important to make the RFP as accurate and as thorough as possible before hiring the CRO. However, if this is the first project the sponsor has put out to bid, it is possible there may be some lapses or mistakes. I would suggest that, following the award of the contract, there be another face-to-face meeting to fully scope out the project. In that meeting, it is possible that some assumptions regarding end products or outcomes were made that weren't adequately addressed in the RFP. It might be painful, but get it right at the beginning. The key point is to have a realistic product development scope and budget and to understand that budget vs. the available sources of funding.

Typically, when budgeting for any project, quality, time and price are the key considerations; under all but the most exceptional circumstances, you usually can't have all three — pick out which of the three is / are most important. Live with your choice!

As much as possible, out-of-scope situations should be avoided. However, if it is clear that additional resources are required for successful completion of the project, acknowledge the fact and acknowledge that an increase in charges will result. On the other hand, careful and thoughtful planning should keep such requests to the minimum.

12.2.4.1 Budgeting

Let's work through budgeting in a bit more detail, starting with terminology.

Fixed fees are independent of study activities. These are assessed no matter what happens in the study. For example, a set up fee for electronic laboratory data processing.

Table 12-8 provides examples, in a tabular form, of costs that are *fixed* during a study. The budgeting might include an initial "engagement" cost. The contractor is being paid to dedicate themselves to the project. Typically for the items listed here, there is still a variable nature to the costs. For example, while there might be an initial fee for the engagement of the project manager, costs associated with her activities might be allocated on a monthly basis and billed for months engaged on the project.

Table 12-8. CRO Budgets: Examples of Fixed Fees

Position	Assumption	% FTE	Time	Rate	Total
Project Manager	For single study, PM is 50–100% throughout entire study.	100%			
Lead CRA	Coordinates monitoring activities during study start-up and conduct periods. Does not include time for monitoring visits.	100%			
Regulatory	For single study, 1Reg at 50–100% during entire study.	100%			
Contracts Manager	Manage investigator grant payments. 1 Mgr at 25% during study conduct.	25%			
IRB Fee	Central IRB fee paid at beginning of study, based on number of sites.				
Total Fixed Costs					

Variable fees are directly dependent upon study activities — the more of them used, the higher the cost. Examples include shipping fees, chest x-rays, each laboratory test performed, etc.

Table 12-9 illustrates some examples of *variable* costs. For example, however many pre-study visits are required for the study are costed on a per-visit basis. The same would be true for other visits listed here and for, as we mentioned previously, individual testing procedures.

Pass-through costs are those incurred by the contractor and billed directly to the sponsor. (As these costs are discussed, determine whether or not there is a markup [it might be described as a processing fee]; if there is, negotiate it.) Also, ensure there is a "not to exceed without prior approval" statement in the contracts.

Next, Table 12-10 identifies a few of the *pass-through costs* that might be incurred in a trial. Examples include CRF printing, shipping and travel costs. Again, understand whether or not any mark-ups are involved. If so, again, negotiate them.

Table 12-9. CRO Budgets: Examples of Variable Fees

Activity	Assumption	# Hours / Visit*	# Visits	Rate / Hour	Total
Pre-Study Visit (PSV)	60% of Sites require PSVs, per SOPs.	32			
Site Initiation Visits (SIV)	Investigator Meeting serves as SIV, per SOPs. 20% of sites require separate SIV.	40			
Routine Monitoring Visits (RMV)	Assume 1 visit after 1st patient enrolled; every 3–4 weeks thereafter.	40			
Close-Out Visits (COV)	100% of sites require COV.	32			
Total Variable Costs					

Table 12-10. CRO Budgets: Examples of Pass-Through Costs

Activity	Assumption	# Units	Rate / Unit	Total
CRF Printing	Assume $250 per CRF book. Assume 1 CRF book per subject.			
Shipping CRFs to sites	Assume 1 initial shipment to Germany of CRFs for distribution to sites at $200 per shipment.			
Travel, Pre-Study Site Visits	Assume $1,200 per trip. Assume regional monitors. (Travel expenses only)			
Travel, Site Initiation Visits	Assume $1,200 per trip. Assume regional monitors. (Travel expenses only)			
Travel, Monitoring Visits	Assume $1,200 per trip. Assume regional monitors. (Travel expenses only)			
Total Pass-Through Costs				

In the final budget, all the fixed, variable and pass-through costs are collected and totaled. (Figure 12-11) Frequently, the budget actually agreed to between the sponsor and the CRO will break out costs by activity group rather than by fixed and variable (but the calculations should be included in the budget, so charges would be quickly understood). The pass-through costs are always identified separately and broken out by item.

Table 12-11. CRO Budgets: Total Costs

Line Item	Total
Total Fixed Costs	
Total Variable Costs	
Total Pass-Through Costs	
Total CRO Costs	$$$

12.2.4.2 Managing the Service Provider

Working with a service provider is very similar to building a house and functioning as the general contractor for the construction of the house.

Consider the question: Would you let a general contractor and his or her subcontractors build your new house while you're on a six-month vacation, hoping that as soon as you come back, your house will be completely ready and set to move into? I think you probably wouldn't want to do that. That's not what should happen here either. The sponsor's project manager needs to stay close to the provider; it's not as though the provider can't be trusted, but the concept of "the squeaky wheel gets the grease" certainly applies here. If the sponsor stays on top of the provider, pays attention, understands the progress, looks at the progress reports, maintains contact with them, the provider will provide better service.

Understand the product development process in detail; that, in turn, enables better management of any contractor or provider. As noted previously, establishing and monitoring relevant metrics will significantly increase the likelihood of a successful completion of the project.

Frequently, the sponsor comes at the project with its own science, vision, and sense of urgency. The sponsor also comes with the money and the oversight. In turn, the CRO provides project teams, trial logistics, specialists, regulatory expertise (especially with multiple dossiers and multiple regulatory agencies), and multilingual worldwide experience. (Of course, "CRO" could equal multiple CROs or individuals.)

When your contribution and the CRO's contribution meet and meld together, the first thought to come to mind might be that perhaps the sponsor has lost control. This is likely to be the thorniest issue. I think that, rather than losing control, the sponsor has employed leverage. Realizing it has limited resources, the sponsor essentially gains a new team. With proper management, control is maintained.

The answer, ultimately, is to integrate the team — to integrate the CRO's and the sponsor's respective groups into a single, focused, well-functioning team committed to project success. Figure 12-3 illustrates the successful combination of resources.

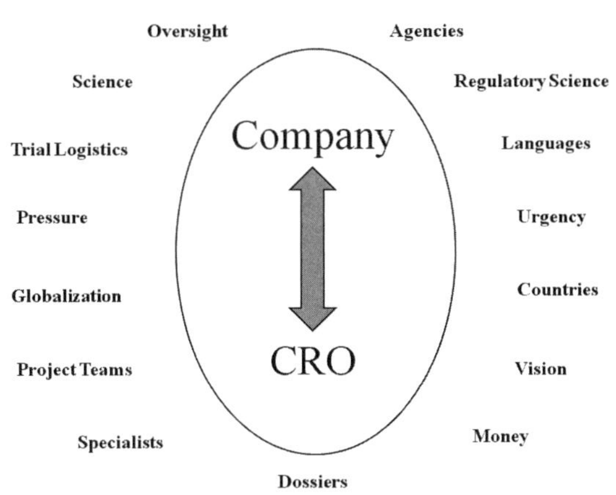

Figure 12-3. Integrated Contributions: Sponsor and CRO

Communication between the sponsor and the CRO or other contractors must be focused and it must be continuous. This communication — to be effective — must be open, honest and complete. Complete means that both good and bad information is conveyed in all its beauty and in all its ugliness. For any team to be successful, the team needs to understand the bad news as it really is and to be able to deal with it. If "bad" isn't heard and the story is incomplete, "bad" can't be dealt with.

The information must be communicated in a way that fosters cooperation, as well as appropriate and reasonable decision-making. In this case, information refers to both the content — truth, good and bad, and the format — email, teleconference, letter, etc. Whatever format is most effective should be employed. Communication is the key to any successful project. It's an important key to any successful outsourcing process.

In working through your relationship(s) with your vendors, develop good habits. Table 12-11 illustrates some habits associated with success.

Table 12-11. Vendor-CRO Relationships: Successful Habits

Establish a strong partnership.
Make the CRO a member of your team.
Establish senior management buy-in.
Function with a sense of urgency.
Don't change in mid-stream.
Always listen.
Never lose control.

An important place to start is developing a strong partnership. Too many people look at contractors — from individuals to multi-nationals CROs — as the enemy. Years ago, when I worked at PAREXEL, we had a very large pharmaceutical company as a client. After the contract was agreed and the project had been initiated, a new person joined the team on their side. As he joined, it wasn't clear what his function was, since he worked in project management and there was already a person with the title of project manager who was staying on the project. After a few weeks, it became clear. His sole function seemed to be — at least to us — to complain about service and price; eventually, his function was to re-negotiate (actually closer to setting a lower price and dictating that it was now in effect). We actually, despite all the harassment, ultimately delivered the product to the *market* for them in a time shorter than originally agreed. Unfortunately, we took a bath on the costs. By the way, PAREXEL hasn't worked with them again.

Another strange affair for your consideration. I was engaged by a biotechnology company not too long ago. The focus of the consulting was first strategic, then tactical — designing and implementing a development program for their lead product outside the US. Eventually, I realized that the CEO (who hired me) and the CFO looked at all contractors — regardless of size — as adversaries. As an example, a CRO in the US executed the clinical trial superbly, enrollment exceeded expectations and the study from start to finish beat the time lines agreed to by nearly six weeks. Yet, at the end of the project, the CEO and CFO latched onto a perceived (not real) misstep and attempted to reduce their final payment by nearly $1 million on a $3.5 million study. Needless to say, no contractor would work with them a second time. By the way, the CEO and CFO were eventually fired — although, in my opinion, a bit late, since the company was headed downward by that point.

These two examples illustrate less-than-optimal methods of developing effective business relationships. I would suggest making the CRO or contractor a member of the team. They really are not simply contractors, they are not simply providers, but (particularly if there is only a single product in development), they are really a crucial core part of the organization. The CRO needs to be treated and worked with as a member of the team.

Be careful of making changes mid-stream. First of all, don't change details of the project unless absolutely necessary. Secondly, don't change the CRO unless absolutely necessary. Clearly, there will be points in time where you just made a bad choice and this CRO just isn't as competent at the task for which they were hired as they were supposed to be. However, don't change unless absolutely necessary. (Usually, changes are needed because you didn't do a good job of scoping out the project at the beginning.)

Assuming the sponsor selected the CRO or other vendors wisely, it is crucial that the sponsor listen to the vendor's experience and expertise. The sponsor might still have a better idea, but completely ignoring the vendor's experience and expertise is a decision fraught with hazard.

Table 12-12 identifies some factors associated with successful CRO or vendor management.

Table 12-12. Factors Associated with Successful CRO Management

Your Company	**CRO**
Vision	Competence
Delegation	Responsiveness
Involvement	Care
Communication	Communication
Accessibility	Accessibility
Sense of urgency	Sense of urgency
Openness	Reputation / image
Steadiness of purpose	Access to complementary capabilities

The sponsor needs to engage each vendor or CRO as partners on the project. Almost always, the results will speak for themselves. Project overhead for project coordination will be reduced. The creativity of the providers will be unleashed and they will enthusiastically "step up" for the team. The sponsor is able to capitalize on the provider's experience and, in turn, augment its own strategic vision and perspective. In the ideal scenario, everyone wins. If the project is a rousing success (even if the product itself were to fail), reward the team members — including the providers.

With great diligence and persistence (and a little luck), the perfect providers have been identified. With adept management and effective communications, the respective teams have melded together as one, all involved are intensely committed to the project, and everyone, at the end, considers the project a success. Life is wonderful and we continue on the pathway to delivering new health care products to those in need.

12.3 Regulation of Contract Service Providers

A short, final sobering note: Contract service providers are subject to regulations. Particular areas of regulation include GMPs, GLPs, and GCPs. In the United States, these responsibilities are covered in 21 CFR 312.52, the section of the Federal Regulations covering transfer of obligations. ICH deals with the issues in E6 in sections 5.2.1 and 5.2.2.

In the end, though, the sponsor is ultimately responsible for everything — so the buck stops with the sponsor. How's that for a heavy burden?

12.4 Summary

In the beginning . . . Understand the objectives of the CRO relationship, know the organization's abilities, be prepared to communicate in a collaborative manner, and (always) ask questions.

In the middle . . . Remember this is a business with the goal of delivering quality products to patients who need them. That's why we're all here. As is true with any project, be it simple or complicated, always expect the unexpected. Anticipate and prepare; put together alternative plans for the things that might go wrong.

In the end . . . Review the process that has been put together. Employ sets of metrics so the team understands success and failure. Evaluate all parties: "How did I do?" "How well did the vendor(s) do?" Regardless of the outcomes (good or bad), assemble that information, analyze it, and determine to do it better the next time.

Final words . . . Don't use a contract service provider if business goals can be achieved faster and more efficiently alone. If you hire a provider, figure out how to balance loss of control vs. intense involvement. Adept and engaged management of, and effective and continuous communication with, the service providers significantly enhances the likelihood of project success.

Part III

Regulating the Plan: Deliver the Best Possible Product to the Customer

Chapter 13
Regulatory: Strategic Considerations

This chapter will address the implications of regulatory strategy for clinical development. Specifically, the chapter will touch on global development, on various strategic options, and considerations of specific populations (e.g., pediatrics) in clinical development.

13.1 Overview

In case this wasn't obvious before . . . drug development occurs in a very highly regulated industry. There seems to be no end to the various agencies (and acronyms) involved — regardless of which country or region you are in.

In Section 1.3.4, we looked at the various contexts, including regulatory, within which development is conducted. While the specific goals of each of the major players in the process — regulatory agencies and sponsors — are different, each player ultimately strives to deliver safe and effective products to patients who need them.

To achieve regulatory success — delivering a product to the marketplace — I believe it is important to establish a collegial relationship between industry and whatever regulatory authority we happen to be working with, in the US, Europe, Japan, or any other regulatory body throughout the world. Those relationships need to be based on mutual respect. (In contrast to some opinions, I don't believe that having a collegial, respectful relationship in any way compromises the regulator's function or the sponsor's principles.)

Incorporated in this viewpoint is a belief that industry must consider the regulator as both customer and supplier. In order for them to supply us with marketing approval, we must satisfy their needs as a customer of the information we provide. We also understand that, in order to be successful, whatever we do has to be based upon a foundation of good science. Sometimes that science is chemistry. Sometimes it's biology. Sometimes it's primarily engineering. But whatever it is, it's a foundation of good quality science. It is also crucial, of course, that industry operates on a foundation of scrupulous honesty.

Regulatory success, likewise, depends upon having a clear understanding of what our customer, in this case a reviewer for a regulatory agency, really wants. We need clarity around that understanding. We can do good science. We can do good quality studies. If we put the information we've gathered from those studies into a format and into a set of data presentations that are incomprehensible, however, we've not dealt well with our customer — the regulatory authority. As a result of the inattention to our first customer, the regulator, our ultimate customer, the patient, does not have access to our product.

Therefore, the deliverable — the regulatory submission — must be clear, concise, convincing and compelling. The reviewer who's reading this submission needs to understand that the sponsor is relating a story with a whole set of characters, plots and developments. The reviewer should feel compelled to continue reading the submission — as with a great story — because of the data presented, because of the conclusions reached. I previously used the term "sales document" to refer to the summaries at the beginning of the CTD. This is the point at which the sales pitch kicks in. However, it can't be style without substance.

Regardless of the market, there is a substantial regulatory hierarchy. In the US, there are very specific components in the regulatory framework: statutes, regulations and guidances.

Statutes are laws passed by acts of Congress or the Legislative Branch. The President, the Executive Branch, signs them into effect. The courts, the Judicial Branch, enforce them. So, they in fact involve all three branches of the US government.

There are also **regulations**, which are rules governing the implementation of the law. They are promulgated by a federal agency, for example the Food and Drug Administration. They do have the full force and effect of law.

In addition, there are **guidances**, for example the Food and Drug Administration guidances or the ICH guidances. Guidances are sanctioned means of implementing regulation. They are not prescriptions about the specific means of doing something, but they are certainly more than suggestions. Whoever has taken the time to create and publish the guidance has a belief that the information is the best available at the time. Accordingly, industry must read, understand and consider the recommendations in the guidance. If the sponsor believes there are alternatives to accomplishing the same goal addressed in the guidance, it is the sponsor's responsibility to discuss those alternatives with the appropriate regulator and gain *explicit* agreement prior to embarking on the alternative pathways.

I do think that there are times when a guidance may not be up-to-date enough to accommodate the particular technology or the indication in which the sponsor is working. Nevertheless, if the sponsor believes an alternative is a reasonable consideration or, perhaps, is the best way to achieve the goal, discussion with the relevant regulator is necessary.

There are guidelines for reviewers (e.g., in US Manuals of Policies and Procedures [MAPPs], for drugs or Blue Book memoranda for devices). These are guidelines for the reviewers. It's important, for those available and applicable to the sponsor's particular area of interest, that the reviewer's approach to submissions be understood. Ignore them at your peril.

There are also industry newsletters, depending on what color you want to pick — Pink Sheet, Grey Sheet, Gold Sheet, etc. Frequently, information in those publications indicates directions in which various regulatory agencies might be moving. The regulatory agencies are fully mindful of the readership of those particular newsletters. There are times when agencies actively use these publications as vehicles of communication.

In Section 3.3, we discussed using approval summaries (US: Summary Basis of Approval, SBA; Europe: European Product Approval Review, EPAR) to assist in the process of assembling a development plan. As a reminder, these documents are put together at the time of product approval (actually, they are released somewhat after approval) of a particular drug or device, chronicling and analyzing the studies performed — non-clinical and clinical — that support the product's approval. The summaries also contain commentary and reaction by the reviewers. It's important to consult these, particularly if the sponsor happens to be working on the second or the third product in a particular therapeutic indication or a particular niche within a therapeutic indication, to understand how another product was able to satisfy the regulatory requirements and be approved for marketing.

In the "old days", the SBA was essentially cut and pasted and put together. There were certain areas in which those old documents contained extremely useful information for development program design. For example, Study X, included in the sponsor's submission, would be described, including the purpose, specific endpoints being studied, and the analyses that were done. The information might also have indicated the number of centers enrolling patients, the names and locations of those centers, and the enrollment rates. Such information is quite valuable in the trial planning process and for the overall design and execution of development plans.

Current SBAs are still cut and pasted, but the computer enables this to be done in a much more efficient manner. Generally, they are now more readable, but some useful information, such as the list of centers, investigators, enrollment rates, etc. is not consistently included. On the other hand, there is generally a somewhat longer discussion of the FDA's interpretation and interaction, so to speak, with the data they are reviewing, as well as the approach the sponsor took in putting the development program together and in analyzing the data. Again, that is very valuable information.

If the sponsor happens to be first in a particular indication or the product is particularly novel, specifically-applicable guidances might not be available. Nevertheless, for SBAs in therapeutic areas somewhat adjacent to the desired indication, or for similar products going through the same review, division may provide useful insights.

In the EU, similar scientific discussions are contained in EPARs. EPARs are generally more succinct (translation: much shorter) than the comparable FDA documents, but they provide highly useful information regarding the EMEA's considerations when reviewing a project.

A lot of variables affect the strategic approach to pharmaceutical development.

Drugs are different. For example, each drug, even structurally similar drugs, has a somewhat different interaction with the disease for which it is being used compared to other drugs.

Sponsors and their organizations are different. Different sponsors and different organizations vary in their approach to drug development, interactions with agencies, and internal processes. The way in which Merck, for example, deals with regulatory agencies is different than the way in which GlaxoSmithKline deals with regulatory agencies. Those variations are partly cultural and partly related to personalities. Furthermore, the ways in which both Merck and GlaxoSmithKline interact with different agencies around the world are likely to be different than the interactions smaller companies might have with exactly the same reviewers. Perhaps the sponsor's organization has several individuals with prior experience, but the organization itself has never had these interactions with regulatory agencies previously. Similarly, it would be hard to argue that the regulatory playing field is level for all players.

Regulatory agencies are different. There are obvious differences from country to country, inter-agency differences. However, even intra-agency differences exist. For example, different FDA centers have different personalities. Different divisions within the centers have different personalities. In addition, obviously, individuals are different. So, the way two reviewers in the same division look at a submission is likely to be somewhat different. This is not to imply that reviewers and divisions don't follow the established procedures, but any individual reviewer will bring different elements of his or her background into the review process — whatever their training it is, whatever their clinical experience might be, whatever the last submission is that they might have reviewed. All of those things introduce variability.

Regulators' attitudes change. There is evolution in the regulatory thinking about risk and benefit. Some of the change is related to changes in science and some to changes in political and cultural thinking. As an example, the FDA has undergone substantial changes in their approach to risk versus benefit. Earlier in the book, we noted the different responses to the AIDS crisis and to the Vioxx situation. I believe user fees have changed the attitude within the FDA, for example, toward increased customer service, but it is nevertheless against a background of heightened sensitivity to risk issues.

Then, there is the issue of time. One of the biggest challenges in assembling development programs that are going to last some period of time — whether it's two years, five years, seven years — is the evolving regulatory climate and potential shift in Agency policies dependent on political expediencies, as well as new scientific information that comes out during the course of that two, five, or seven years development plan we are in the process of executing.

All of these elements introduce variability into the regulatory process that we need to pay attention to and to be mindful of as we put together our development programs.

Certainly, the international harmonization process (the ICH process) has introduced a more common (not identical) playing field across at least the three major geographic areas in the world. The ICH process has also provided a frame of reference now used by regulatory agencies in many other regions and countries around the world. Some countries have essentially adopted the ICH process; other countries use the ICH process in concert with their own regulations. In any case, drug development around the world has been altered by ICH.

Before heading into the discussion of global registration strategies, I will offer a couple of philosophical thoughts for the reader's consideration. First, someone told me years ago — and I have used it as a guiding principle since — "Regulatory strategy is a competitive weapon." A corollary to that statement is the comment that, although regulations are written in black and white, there is a lot of grey in them, in whatever regulatory jurisdiction one operates. Given there is grey in the regulations, it makes sense to exploit the greyness. The greyness is a ticket to be thoughtful and creative (never deceitful or dishonest).

The second comment is to always keep in mind that approval for marketing and reimbursement for the marketed product are not the same things. For example, it is common for US device companies to assume that receiving a CE mark equals reimbursement — it doesn't. All countries — US included — have reimbursement policies that are separate from market approval policies. Optimally, in the development plan, those activities geared primarily to registration and marketing approval and those activities geared primarily toward reimbursement are integrated into a single, effective plan that achieves both objectives.

13.2 Global Registration

One of the terrific benefits of the move to the common technical document (CTD) is the ability to develop what is essentially a single dossier for the world. Although there are regionally-specific portions of the submission, the essence of the CTD is accepted world-wide (even in many non-ICH regions).

Figure 13-1 depicts the organization of the CTD, building on the foundations of Quality (the CMC section), Safety (the non-clinical sections), and Efficacy (the clinical sections), and moving through progressively more integrated summaries. It is built from the specific to the general, from data to labeling.

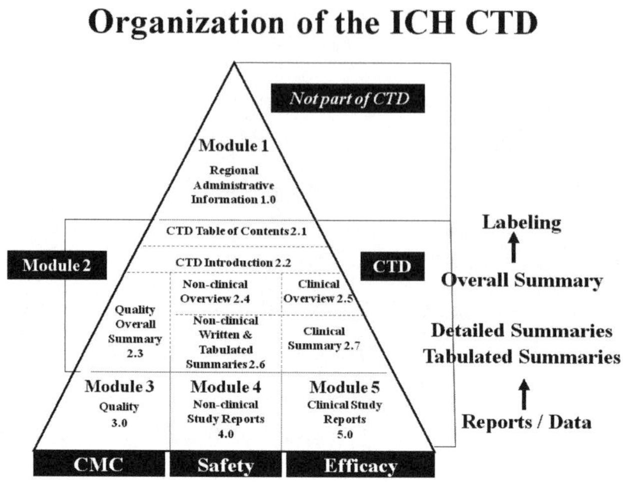

Figure 13-1. Organization of the CTD

Figure 13-2 illustrates the integration of the disease process-development pathway with the appropriate sections of the CTD. Starting with the disease process addressed by the product, moving through definition and specification of a drug candidate and proceeding through safety and efficacy evaluations, first in animals and then in man, the sponsor can "map" all of the information generated into the designated section of the CTD.

Benefits to industry of the CTD include the common structure, common content, and common file format. With a little practice it is relatively easy to produce and to organize. (With a little mapping help, it is relatively straightforward to move from the old system to the CTD.)

Regulatory agencies also see benefit in the CTD format. The CTD identifies the correct content for the reviewing agencies in an organizational format that is consistent from beginning to end (i.e., from initial filing through market approval application). The regulatory agency can pull it apart and review it in accordance with its own procedures and preferences. The format also enables both the sponsor and the reviewing agency to manage the life cycle of the product's development in an integrated fashion.

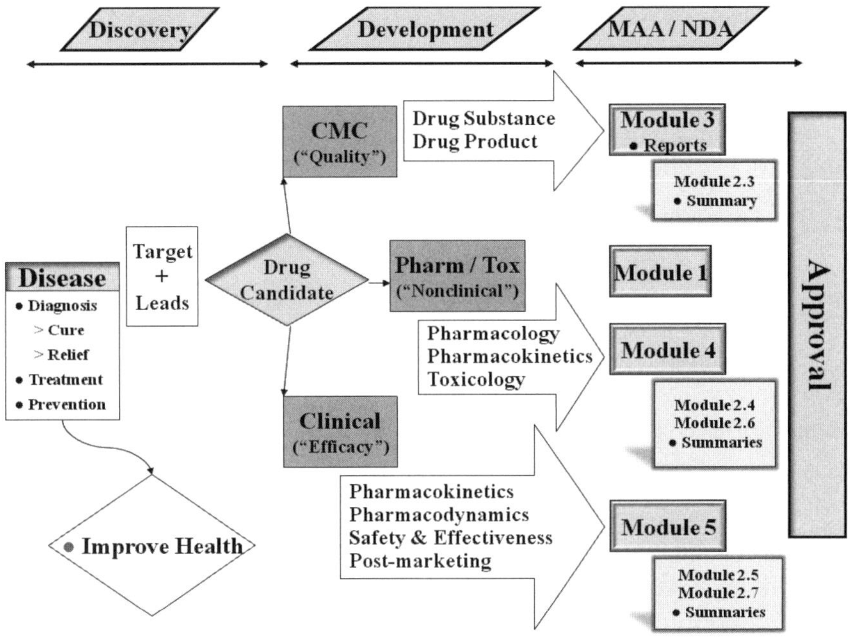

Figure 13-2. Integration of Development Processes and the CTD

While a step forward, the CTD format has some issues that will need to be ironed out over the next few years. For example, different agencies do review it differently. Agencies have different experiences. An example, mentioned in Section 10.1.2, is the large difference in page counts for the same sections in CTD submissions to the FDA.

One really critical area that does need active attention is information technology (IT) support at the various agencies. Many regulatory agencies operate with a patchwork system that has been assembled piecemeal over the past several decades. Not all agencies have the same technical capabilities — particularly outside the core of the ICH countries (but even within ICH countries).

Lastly, since the whole process is an evolutionary one, different agencies have different ideas of the future. The sponsor's job is to understand the differences and to manage them as the sponsor brings its products to market.

In summary, I would suggest the CTD is a (very) good vehicle to achieve a goal. It alone is not the problem or the solution. Ultimately, the CTD is only as good as the data, its authors, and the product it represents. So design well, execute superbly, pray a lot and it will all make sense.

13.3 Strategic Considerations — Where? Why? When?

Next, let's consider some strategic options available for the development program. (See Table 13-1.)

Table 13-1. Regulatory Strategic Options

Regulatory Strategic Options
Orphan Drug
Humanitarian Device Exemption
Expedited Development
Fast Track
Subpart E
Subpart H
Rolling Review
Priority Review
Early Access
Treatment IND
Parallel Track
Exploratory IND
Special Protocol Assessment

13.3.1 Orphan Designation

13.3.1.1 United States

In the United States an orphan drug application can be submitted any time between IND filing and the time an NDA is filed. It is submitted to the FDA's Office of Orphan Product Development (OOPD). By statute, OOPD has 60 days to review the submission. Since OOPD is fairly efficient, that 60-day decision time frame is pretty real.

The determination of orphan status is based primarily on demographic criteria. An affirmative decision at this point confers orphan status on the product. Criteria in the US are EITHER there are fewer than 200,000 people in the US with the diagnosis OR there is no "reasonable expectation" that costs for development of the drug to treat a disease can be recovered by sales of that drug in the US.

With regard to the definition of the population, some sponsors have taken a diagnosis and continued to subset it until they got to a group that was less than 200,000. That exercise is useful if the population is still a meaningful group from a diagnostic standpoint and reasonable criteria exist to provide that diagnosis.

Table 13-2. Advantages to Sponsor of Receiving Orphan Designation

Tax credit for development costs — can carry forward
Eligibility to apply for an orphan product grant
Protocol development assistance from the Review Division
OOPD assistance with the Review Division through an ombudsman
No user fee for NDA or BLA review
Seven years of market exclusivity after approval
High potential for priority review of the marketing application
On average, takes five years from designation to marketing authorization in the US

Table 13-2 illustrates some of the advantages to the sponsor of obtaining an orphan designation. The advantages for the sponsor include tax credits for development costs that can carry forward. There is eligibility to apply for an orphan product grant for development. The grants can be meaningful in terms of size, but must be for future work (not previously run studies), and applications are accepted only once a year.

There is protocol assistance from the reviewing division, as well as the assistance from the Office of Orphan Product Development. Also, there is no user fee for an NDA / BLA filed — that is a significant chunk of change. Reviews of orphan applications tend to go faster, and there is the carrot of seven years of market exclusivity available.

In addition, as many working in smaller companies know, whenever a company makes an announcement that says, "We now have orphan drug status," it is, in general, favorably regarded within the investment community.

Orphan designation really is a horse race. The designation is granted to the first drug with orphan status to be approved in the particular orphan indication. The challenge for orphan designation is that the designation is for a specific chemical compound. So, if a sponsor is developing something for cystic fibrosis and another company has a different compound in development for cystic fibrosis that may have a somewhat similar mode of action, but it's a different compound, that drug can also achieve orphan drug status. Thus, several sponsors may receive an orphan designation at the end of their respective development processes.

Orphan designation is valuable from a financial standpoint because of some of the incentives built in. It is valuable from a regulatory standpoint because there is direct assistance from the Agency. On the other hand, keep in mind what that exclusivity really means: It applies to the specific compound, not to the disease entity.

Table 13-3 provides a list of some starting resources regarding orphan development and designation in the US.

Table 13-3. Resources for Orphan Development — US

FDA, Office of Orphan Products Development (OOPD)
www.fda.gov/orphan
National Organization for Rare Disorders
www.rarediseases.org
NIH Office of Rare Diseases (ORD)
www.rarediseases.info.nih.gov

13.3.1.2 European Union

The EU also has an orphan drug process. It is the result of the EU Regulation on Orphan Medical Products that was implemented in 2000. It comes under the purview of the Committee for Orphan Medicinal Products (COMP).

In the EU there are also two criteria: EITHER the disease affects no more than 5 / 10,000 persons in the EU at the time of submission of the application for designation, OR, similar to the US, the sales are unlikely to cover development and there is nothing currently available for the seriously debilitating condition in which we are interested.

Likewise, there are regulatory and marketing benefits. COMP offers protocol assistance and there are fee reductions / exemptions for marketing applications. Following approval, there is a 10-year marketing exclusivity provision.

Table 13-4 lists resources regarding orphan designation in the EU.

Table 13-4. Resources for Orphan Development — EU

European Commission Directorate-General for Health and Consumer Protection
www.ec.europa.eu/health/ph_treats/non_com/rare_diseases_en.htm
EUORIS (umbrella organization of patients' associations in EU)
http://www.eurordis.org/
Orphanet (also location of collected resources in EU)
www.orpha.net

13.3.2 Humanitarian Device Exemption

The humanitarian device exemption applies to devices used in fewer than 4,000 patients per year in the United States. The approval of the application permits the sale of the device in the absence of either 510K clearance or PMA approval. In other words, it permits use and sale of the device in the absence of conclusive evidence of safety or effectiveness. The indication for which the device can be used must be for serious or life-threatening conditions for which there is no comparable therapy approval. Also, the device may only be used in facilities that have IRBs to supervise the clinical testing and treatment.

The relevant CDRH guidance is located at http://www.fda.gov/MedicalDevices/DeviceRegulationandGuidance/HowtoMarketYourDevice/PremarketSubmissions/HumanitarianDeviceExemption/default.htm.

13.3.3 Expedited Development

There are various processes for expediting the approval of drugs.

13.3.3.1 Fast Track

The key question in considering the Fast Track process is: "Is the drug to be used in a serious or life threatening condition for which there is no comparable or adequate therapy currently approved?" If so, the drug may be eligible for fast track designation.

This was codified in the FDAMA legislation in 1997. It enables more frequent contact with the FDA and provides an option for submitting an NDA in sections — what is called a "rolling review". It also presents the option of requesting evaluation based on surrogate endpoints. It is a process independent of priority review and accelerated approval.

The Guidance, revised in 2006, is located at http://www.fda.gov/downloads/Drugs/GuidanceComplianceRegulatoryInformation/Guidances/ucm079736.pdf.

It is important to understand that consideration of Fast Track designation applies to a combination of *product + claim* that addresses an unmet need. As mentioned previously, an essential criterion is that the product be intended for treatment of a serious or life-threatening condition and data regarding use of the product, to date, demonstrate the potential to address unmet needs for that medical condition.

13.3.3.2 Rolling Review

A rolling review entails submission of sections of the NDA or BLA prior to submission of the full document. There are advantages to this, particularly if the CMC work (other than final stability data on the three manufacturing lots) is essentially complete. The CMC section is generally the first section submitted in this scenario. However, the pharmacology / toxicology section also can be submitted early if data are complete. Many of these types of programs will not have any, or will have only incomplete carcinogenicity data at the time of submission anyway, so those data wouldn't be included in this scenario.

Technically, the FDA isn't obligated to review the sections until the full NDA is received. However, for orphan diseases and serious diseases with unmet needs, the Agency is trying to assist the sponsor in moving the product forward, so the Agency has a vested interest in an early review.

Noteworthy, however, is the fact that the user fees are due with the submission of the first section. Bummer!

13.3.3.3 Subpart E

Subpart E is covered in 21CFR 312.80 in an older section (1988) and addresses *expedited development*. Subpart E is a process that is intended to expedite "the development, evaluation, and marketing of new therapies intended to treat persons with life-threatening and severely-debilitating illnesses, especially where no satisfactory alternative therapy exists".

The designation enables early and frequent FDA interactions that are intended to compress Phase 2 and Phase 3. Generally, the approval is based on a smaller total safety data package, but always includes a Phase IV commitment.

13.3.3.4 Subpart H

Subpart H addresses *accelerated approval*. The relevant regulations are located at 21CFR 314.

In the Subpart H process, the product may receive a conditional approval before the primary endpoints of the pivotal study or program are achieved. For example, mortality and morbidity data may not be completely available, but a set of surrogate endpoints have been met, suggesting benefit to morbidity and mortality. The FDA would then proceed with the conditional approval. The condition is that Phase 4 studies *will* be conducted to confirm the efficacy end point or, perhaps simply, that studies already underway will be completed to evaluate morbidity and mortality.

To recapitulate, Subpart E represents expedited development, and Subpart H represents accelerated approval.

13.3.3.5 Priority Review

Priority review represents a designation separate from the other fast track designations.

Following submission of a marketing application (NDA), the appropriate division will review the submission and determine the priority. The designation — standard or priority — determines the review timing.

For a standard designation, the target is for an action (approved / not approved) to occur within 10 months of submission The target for review of a product with a priority designation is six months. Clearly, these are targets rather than firm commitments. Nevertheless, all other things being equal, you would prefer to have a priority designation.

13.3.4 Early Access Programs

Early access programs include treatment INDs, parallel track programs, and making an investigational product available to individuals NOT enrolled in a controlled clinical trial.

13.3.4.1 Treatment IND

Treatment INDs really represent a regulatory compromise. From one point of view, patients who are desperately ill should theoretically have some access to emerging therapy. Certainly, this is easiest to understand in areas of oncology, for example. Patients have failed two or three or four courses of chemotherapy. They've failed all the standard courses of chemotherapy. You have a product for which this is your target indication. In this situation, though, your product would represent rescue therapy. The likelihood, in fact, of having a response to your particular product is relatively small.

An alternative point of view regarding treatment INDs is that investigational drugs should be withheld until testing is completed and the regulatory agency judges the drug to be safe and effective.

These two conflicting points of view come together when considering a treatment IND.

The treatment IND has its roots in the 60s and 70s. The concept started with "compassionate use". Various antiarrhythmics were used under this program. The problem with antiarrhythmics is that a significant body of work is required to put an anti-arrhythmic into clinical practice. However, many of the products seemed to have efficacy, yet were not approved. Hence, certain patients with significant arrhythmias didn't have access to the latest products. Treatment INDs became one mechanism to address this issue.

Originally, beta blockers were used in the area of hypertension, but we found out early that patients with angina had benefit from use of beta blockers as well. So, treatment INDs were put in place so the patients would have access to the drug in angina as well.

The treatment IND is almost always in the name of the investigator, rather than the sponsor. As a result, the sponsor, at the time of filing for market approval, has to deal with data that may be somewhat chaotic, at best. Treatment INDs were ultimately codified in the FDAMA legislation in 1997.

The treatment IND is used in cases where there is positive preliminary evidence of safety and effectiveness. The argument is that the potential benefits outweigh potential risks. The medical condition is serious with no satisfactory alternative therapy. Treatment IND criteria are listed on Table 13-5.

As an interesting side note, the sponsor can "sell the product" although technically, the product may not be sold at a profit.

Table 13 – 5. Treatment IND Criteria

Intended to treat serious or immediately life-threatening disease
No comparable or satisfactory alternative
Drug is under investigation in a controlled clinical trial under an IND
Sponsor is actively pursuing marketing approval

In general, the investigator puts together a protocol to address a specific subset of the population being addressed in the sponsor's IND. The investigator will also (with the sponsor's permission) reference the sponsor's IND in terms of other CMC, pharmacology / toxicology, and safety information. A sponsor also can submit a treatment protocol to an existing IND. It is, however, more commonly done by the investigator.

Usually, when an investigator submits a treatment IND, as I just mentioned, that submission will cross reference the sponsor's IND. Obviously, in this circumstance there is a lot of cooperation between that investigator and the sponsor. Following a 30-day wait, the investigator or sponsor may proceed with the study, assuming the Agency has not identified any specific objections.

One issue the Agency attempts to understand with the treatment IND is whether it is being used by the sponsor to go around the development process and collect data in a different way. Accordingly, treatment INDs are not particularly encouraged by the Agency.

On the sponsor side, I would suggest it is not beneficial to have a lot of treatment INDs out there, simply because there is less control over the data, the data collections, adverse event reporting and so forth. All of those elements can affect the review of the NDA.

So, while there is some potential benefit — for example, an investigator with a special interest in a particular form of the targeted disease might be able to efficiently conduct an exploratory study and obtain useful preliminary data — there is always concern about adherence to protocol and data collection standards, as well as diligent monitoring of adverse events. Carefully consider all implications before heading down this pathway.

13.3.4.2 Parallel Track Program

The parallel track program was originally initiated in 1992 for AIDS drugs. It has a somewhat more liberal mechanism for product access and, ultimately, product approvals. It was based on an assumption of insufficient effectiveness evidence being available to meet the threshold of a standard IND and standard NDA.

While drug availability is provided after Phase 1, the challenge with this pathway is creating buzz and open-label access while continuing the traditional development program. How are patients allocated? Will enough patients enroll in the trials to enable adequate data collection? Will there be enough comparative data available at the end of the program? Nevertheless, the process has certainly enabled earlier access to critically needed medication, e.g., AIDS drugs.

13.3.4.3 Potential Issues with Early Access Programs

A potential downside of all early access programs is the possible negative effect on enrollment in controlled clinical trials, as noted briefly in the summary of the Parallel Track Program.

An additional issue has to do with reimbursement for activities surrounding a clinical trial. There is no third party reimbursement for this type of investigational drug, in general. It might be sold, but the sponsor is at the mercy of that patient's ability to pay for the particular product! As a reminder, until approval, there is no mechanism for profit. Many insurance policies will have a section that specifically deals with investigational drugs and, in general, they are not covered "without the approval of the medical director of the program". Reimbursement under that condition is certainly not a foregone conclusion.

Despite some of the potential issues, these programs are on the menu of choices for consideration by the sponsor as the development program is assembled.

13.3.5 Exploratory IND

The concept of an Exploratory IND is an outgrowth of the Critical Path Initiative, about which we will spend some more time in Chapter 15. It is really a "Phase 0" study. It represents a phase — in man — occurring after non-clinical studies and before traditional Phase 1 studies. The format enables the use of microdose or low-dose titration periods of drug administration that presumably will produce pharmacologic, but not toxic, effects.

Although I mentioned "after non-clinical studies", it is possible (in fact, preferable) to move to these studies with an abbreviated toxicology package. An abbreviated package would contain data generated by using the intended route of administration in a study that incorporates a minimum of a 14-day observation period. Data include both clinical and histopathological evaluations and the studies need to have been conducted under GLP conditions. However, and this is a major element of the criteria, it is not necessary to have established a maximally-tolerated dose in animal studies prior to moving to the human study.

The guidance relating to Exploratory INDs, released by the FDA in January 2006, however, very clearly limits human exposure to exploratory studies with NO therapeutic or diagnostic intent. By definition this would limit the number of subjects, the dose range employed, and the period of exposure. The challenge is to identify the appropriate marker of pharmacologic response. (The guidance is located at http://www.fda.gov/downloads/Drugs/GuidanceComplianceRegulatoryInformation/Guidances/ucm078933.pdf.)

The utility of this approach is to demonstrate proof-of-mechanism while potentially performing dose ranging (dose response) relative to that pharmacologic marker. Alternatively, the technique could be used to screen multiple compounds in the same setting. In the right circumstance, with the right markers and under the proper scientific conditions, microdose exploratory INDs offer significant potential for accelerating the drug screening and drug development process.

13.3.6 Special Protocol Assessment

The Special Protocol Assessment (SPA) was created by FDAMA in 1997. An official guidance was released in 2002. (http://www.fda.gov/downloads/Drugs/GuidanceComplianceRegulatoryInformation/Guidances/ucm080571.pdf).

In general, there are three types of protocols eligible for consideration: First, animal carcinogenicity protocols; second, drug product (particularly biologic) stability protocols; and third, clinical protocols.

With regard to clinical protocols, the SPA assignation applies only to studies forming the primary basis for efficacy claims in the marketing application. Most commonly, the protocols address an endpoint that is "less than standard". It is an endpoint that perhaps the FDA has not used previously for approval, and in the SPA the endpoints are very meticulously defined and specified.

The FDA will evaluate the SPA application within 45 days of submission and, in their review, the primary question is, "Are the endpoints defined well enough and with enough statistical power to be adequate for scientific and regulatory requirements?"

In the request for the review, the sponsor poses questions regarding specific aspects of the study such as the objectives, the design, the endpoints, the statistical analysis (including the power calculations) and the definition of the population (such as inclusion and exclusion criteria) and then provides answers suggesting this protocol, as proposed, answers all those questions positively.

The application also lays out the proposed regulatory outcome (i.e., approval based on an NDA containing the following numbers of patients, what the efficacy profile looks like . . . and what the safety profile looks like . . .) and the proposed labeling.

The good news to this process is that the definitions of success are pretty clear. The bad news to the process is that the hurdles are generally set pretty high.

13.4 Population / Demographic Considerations

Next, we will look at different population and demographic considerations, focusing on: geriatrics, pediatrics, and women of child-bearing potential.

13.4.1 Geriatrics

I am particularly interested in this section since I am rapidly approaching this category! It seems intuitively obvious that if a sponsor believes its drug will be used in the geriatric population, that it would be tested in the geriatric population.

As an example, antihypertensive drugs will be used in patients who are elderly. Thus, programs addressing this indication need to include studies that specifically address an older population. However, for years, the industry developed antihypertensive drugs in the 21- to 59-year-old population and then used them in any patient still breathing.

I would suggest that the foundation of any approach to the geriatric area needs to be based on pharmacokinetics and pharmacodynamics, in order to accurately understand dose relationships in the geriatric population.

Currently, understanding what a geriatric population is may be a little bit more problematic, since people who are in their 60s and certainly into their 70s are very healthy. It is certainly different than, for example, 50 years ago when people who got to be that age tended to have a significant number of diseases. Today, people in this age range tend to be active and may have a physiology that's more akin to a 50-year-old than to an 80-year-old. Nevertheless, development programs for such agents need to evaluate the pharmacokinetics and pharmacodynamics in the relevant population so that the effects of the drug, in relationship to the disease, in the context of patients of this age, would be well understood.

Drug-drug interactions, related to other drugs commonly used in the population, are also important to understand, as are drug-disease interactions in older patients. Additionally, the potential incidence of renal disease and, also maybe to a lesser extent, to liver disease in the aged population are important to understand as effects of the drug in the elderly population are explored.

Certainly, based on these comments, it is no longer tenable to arbitrarily exclude geriatric patients from clinical trial development programs.

The location of the ICH guidance is http://www.ema.europa.eu/pdfs/human/ich/037995en.pdf (ICH E7).

13.4.2 Pediatrics

Pediatrics is generally defined as 16 years of age or less, but can also go up to 18.

Historically, pharmaceutical companies have not evaluated drugs in pediatric populations. They've introduced drugs into the marketplace and then pediatricians would begin to use the drugs in the patient population, simply because the drugs appeared to be beneficial for their patients. They might modify doses based on weight or age or body mass. But the process was controlled only in the sense that the pediatrician was paying attention to the product and its effect on his or her own patient. On the other hand, no specific studies were done in those patient populations. In the past, essentially all usage of drugs in the pediatric population was off-label since there was no information in the product labeling related to these patients. While certainly legal, pediatric patients were nevertheless placed at additional risk.

This has now been changed with the requirement for testing in pediatric populations during the course of a development program. ICH has prepared a guidance for this population: http:/www.emea.europa.eu/pdfs/human/ich/271199en.pdf (ICH E 11).

In 1979, the first attempts to include better information in product labels for pediatric usage focused on including pediatric administration in the Indications and Usage section of the drug label. In addition, the appropriate pediatric dosing was to be placed in the Dosage and Administration section. Also, the Agency required a "Pediatric Use" statement be included in the label if safety and efficacy information could not be established for pediatrics or, particularly, if pediatric usage was associated with a particular hazard. Despite these efforts, few products on the market in the early 1990s have labeling information regarding pediatric use.

The first pediatric rule was published in 1994 and encouraged the submission of pediatric data regarding use of a drug, even if "adequate and well-controlled" studies were not available. The Agency allowed extrapolation of adult efficacy data to pediatric application in certain situations, such as similar drug effects in the same disease process. (Of course, one wouldn't know about the latter unless there were some data.) The rule was promulgated with a request for voluntary compliance. Again, the industry response wasn't even poor, it was lousy.

The Agency tried again in 1998 and issued another pediatric rule. In this case, pediatric studies were mandated for all new molecular entities, new indications, new dosage forms, new dosing regimens, and new routes of administration. HOWEVER, the rule also set forth specific conditions for waivers or deferral. The rule also authorized the Agency to require pediatric studies for previously-approved drugs or biologics when the product was determined to be used in a "substantial number" of pediatric patients, or when the product was believed to provide therapeutic benefit in pediatric patients greater than did currently available and labeled products, or when inadequate labeling could pose significant risk.

The FDAMA legislation of 1997 tried another approach — the economic one. It enabled an economic incentive for the submission of pediatric data by providing a six-month extension of marketing exclusivity and patent life following submission and review of data. As a threshold for requiring studies in the pediatric population, two terms came into play: "substantial use" and "meaningful therapeutic benefit".

The definition of **"substantial use"** is use in 50,000 or more pediatric patients with the condition for which the product would be used. If that threshold is hit, then "substantial use" as a criterion comes into play and, as a result, studies must be conducted. An alternative definition focuses on sub-populations. In that case, the definition of "substantial use" is 15,000 or more affected in the designated sub-population. For example, it might be a particular age range or could be a developmental stage in the growth of a child. If the number is achieved within the defined subgroup, that constitutes "substantial use" and requires pediatric studies.

The next criterion for determining the need to conduct pediatric studies is **"meaningful therapeutic benefit"**. This requires evidence of increased effectiveness, elimination or reduction of a treatment-limiting drug reaction. The latter would be represented by a concomitantly taken drug that would ameliorate some of the side effects of the first drug or enable one to use the first drug over a longer period of time. This criterion also applies if the drug enables enhanced compliance or perhaps evidence of safety and effectiveness in a new sub-population or age group.

Some conditions that would enable application for the pediatric development waiver include the following:

- The condition is not relevant in children. For example, since prostate cancer is exceedingly unlikely to occur in a child, a product being developed solely for use in prostate cancer would logically be subject to a waiver.
- The product does not meet the criteria for pediatric populations listed above.
- The studies for logistical reasons might be impossible (or potentially unethical) to conduct in a pediatric population.
- There is substantial evidence that the product would be ineffective or unsafe in a pediatric population.

It is also possible to consider a deferral of studies in a pediatric population until the product has been approved in an older population. This, again, is a matter of negotiation with the Agency. For example, the sponsor believes safety and effectiveness information in adults should be collected first, and obtaining pediatric data would significantly delay availability of the product for adult use. The argument would be that, at a point at which "adequate" data are available regarding safety in the older population, studies for the pediatric population could be designed and implemented. However, because of the "late" start, the sponsor would be requesting the first approval for the adult population to proceed and then to submit the pediatric data when available. The approach makes sense, but needs to be discussed and approved with the relevant Division.

Another logistical challenge might occur as a result of extremely slow enrollment in the pediatric studies. The delay would put the rest of the program at risk and the use in pediatric population would appear to be relatively small. In that case, a waiver request would be a reasonable approach.

To apply for a Pediatric Exclusivity Extension, the sponsor would submit a proposal in a written request to the FDA and file a report on the requested studies. Assuming the sponsor met the requirements, the Agency would issue the pediatric exclusion extension, adding six months of exclusivity after the original patent expiration.

Table 13-6. Resources for Pediatric Development

FDA Amendments Act of 2007 (FDAAA)
www.fda.gov/RegulatoryInformation/Legislation/FederalFoodDrugandCosmeticActFDCAct/SignificantAmendmentstotheFDCAct/FoodandDrugAdministrationAmendmentsActof2007/default.htm
Best Pharmaceuticals for Children Act (BPCA)
http://bpca.nichd.nih.gov/
BPCA / Pediatric Exclusivity
http://www.fda.gov/downloads/Drugs/DevelopmentApprovalProcess/DevelopmentResources/UCM049924.pdf
Pediatric Research Equity Act (PREA)
http://www.thefederalregister.com/d.p/2005-09-07-05-17694
Clinical Investigation of Medicinal Products in the Paediatric Population (ICH E11)
http://www.ema.europa.eu/pdfs/human/ich/271199en.pdf

In Europe, the Paediatric Regulation (Regulation (EC) No 1901 / 2006, updated in 2008) requires submission of a development plan — the Paediatric Investigation Plan (PIP) — that is aimed at ensuring collection of appropriate data in children to support authorization of the medicine for use in children. As in the US, waivers may be obtained if use of the medication in children is not appropriate.

Table 13-6 lists several resources regarding pediatric development.

13.4.3 Women with Child-Bearing Potential

Over the past several years, particularly within the last decade, there has been increased emphasis on providing access to clinical trials to women.

There has always been a concern for sponsors who were developing drugs about the use of women who might be unknowingly pregnant at the start of a woman's participation in a study. In studies it is the rule to write into the exclusion criteria that no woman who could be pregnant could be enrolled. This is generally covered by 1) identifying women who are no longer of reproductive potential (e.g., women past menopause, women surgically sterilized) and 2) assessing pregnancy tests at the start of, and periodically through, the trial. In addition, studies might also specify the use of contraceptive agents to prevent pregnancies in the trial.

However, the Agency is still concerned about the potential for systematic exclusion of women of reproductive potential, even if not pregnant, from participation in some or even most clinical trials. The Agency's position — certainly reasonable — is that this population is very large and, post approval, is likely to be exposed to the product; therefore, it is important to include such individuals in trials. The quandary for sponsors, of course, is to balance the reluctance to expose fetuses or infants to an experimental drug versus the need to obtain experience in an adult female population.

So, the FDA proposed rules authorizing a clinical trial hold if women of reproductive potential were excluded solely because of the risk of developmental or reproductive toxicity. Their primary goal was to ensure that women of reproductive potential with life-threatening conditions were not automatically excluded for that reason alone. It is consistent with the principle of "beneficence" described in Chapter 7.

This chapter has reviewed some strategic topics for consideration as a development plan is assembled.

Chapter 14

Regulatory: Operations

14.1 Meetings with Regulatory Agencies

Meetings with regulatory agencies are important means of communication between sponsor and agency.

Table 14-1 identifies several meetings that might be held between the sponsor and the US FDA. The FDA has provided a general guidance for meeting with them entitled, "Formal Meetings Between the FDA and Sponsors or Applicants", newly revised in May 2009. It would be useful to consult that guidance prior to requesting a meeting with the FDA. The document is located on the FDA web site. (http://www.fda.gov/downloads/Drugs/GuidanceComplianceRegulatoryInformation/Guidances/UCM153222.pdf). The European EMEA likewise has a series of available meetings and consultations available to product sponsors. Consult the EMEA web site. (http://www.emea.europa.eu/htms/human/raguidelines/intro.htm). The discussion below will focus on the FDA meetings, although the suggestions regarding preparation and presentation are applicable to either FDA or EMEA meetings.

Particularly if the sponsor is new to this process, it might be helpful to arrange a strategic planning meeting with the Agency Division to which submissions will be provided. This meeting will occur via teleconference early in the development process. When the meeting is set up, the sponsor should have at least enough information about the product to have identified proposed indication(s) and to have sketched out the proposed development plan. The sponsor might propose options and ask questions regarding the overall regulatory approach for the program. If the sponsor has questions, this is an ideal opportunity to have them answered.

Table 14-1. Meetings with Regulatory Agencies

Stage of Development	Objective	Usual Mode
Strategic Planning	Regulatory Pathway Non-Clinical Testing	Teleconference
Pre-IND	Verify Acceptability Discuss Plans	Teleconference
Phase 1 / 2	Update FDA Review Basis for Approval Clinical Utility	Meeting
Phase 3	Proof — Safety and Efficacy	Meeting
Pre-Filing	Inform / Update FDA Agreement on Filing	Meeting

14.1.1 Overview

In order to have successful meetings with the FDA and other regulatory agencies, there are a few rules and guidelines to pay attention to.

First, and most importantly, define the objective of the meeting itself. "What am I trying to get out of this meeting? What is the agreement I would like to achieve with the Agency? What question(s) would I like to have answered? What would I like to convince the FDA of?" Whatever the objective is, identify it clearly.

Secondly, get help in the planning. If the sponsor is a very large company, for example GlaxoSmithKline, Merck, Pfizer, etc., adequate resources are generally available within the company. But even in those circumstances, occasionally such companies will go outside for external consultants to get feedback and an outside opinion. On the other hand, if the sponsor is a relatively small company and there is not a lot of internal expertise, that external advice is particularly valuable.

In addition, a thorough briefing document, containing essential information (including both good and bad information), should be prepared and submitted to the Agency. Keep in mind, there is no perfect product! The Agency is aware of that.

The information provided in the documents should tell a story and reach a conclusion. It is important to identify the issues to be addressed with the Agency. I believe it is best to also propose resolutions and "force" the Agency to respond to your proposal. Never, never, never simply ask the Agency for open-ended advice. In general, they won't provide it; however, if they venture down that path, the advice received might not really be wanted! Always propose a solution — the best guess based on available data and available advice, or perhaps a couple of alternatives. For whatever options are proposed, the sponsor needs to fully understand the ramifications of the respective choices.

In addition to the document sent to the FDA ahead of the meeting, the sponsor prepares a comprehensive presentation that includes both "core" material and "back up" materials. The core materials are the materials to be presented; the back up information is available in case there are questions best answered with additional materials. Meetings that are conducted under a teleconference format generally operate from the submitted documents rather than presentation materials. Face-to-face meetings are best organized around the presentation materials with the submission documents as the primary support.

It is important that the number of slides to be presented and the time required to present them fit into the time slot allowed. In general, meeting times are 90 minutes. The Agency will NOT extend the meeting. Make sure the slides tell the story, identify the key issues, and propose resolutions. Ensure that the time for presentation includes time for questions and discussion. That will be the most valuable part of the meeting. To accommodate questions and discussion, a good rule of thumb is to allow about 2–3 minutes per slide. In 90 minutes, that would mean approximately 30–40 slides (including titles and transitions!).

It is crucially important that the sponsor develop a positive, collegial relationship with the Agency. For the most part, the vast majority of the people who work at the Agency are interested in public safety and in working with the sponsor. While I stress having a positive and cooperative attitude, it is vitally important to come to a resolution on each issue. I am not suggesting "caving" on all points of disagreement, but make sure the conversation is conducted in a cooperative manner.

As an operational item, identify a team spokesperson who will direct questions posed by the Agency to the appropriate team member. Also, have a team "reporter" who will be taking notes about questions asked and agreements reached.

By the way, practice, practice, practice. Part of the practice is simply going over the information to make sure those people presenting know the material cold. In addition, particularly if the sponsor is a small company, I would suggest obtaining outside assistance at those review sessions from experienced people. Essentially, the goal is to make sure the FDA does not ask a question that hasn't been previously considered. If it is necessary to hire 10 consultants to prepare for all the important areas, do it! Do whatever it necessary to get it right.

In one of my first big pharmaceutical company stops, there was a curmudgeonly regulatory type who served this function; he had an uncanny ability to ask the same questions the FDA was going to ask a week or two later. Sometimes I didn't like his badgering, but, at the end of the process, I was always well prepared for those discussions with the Agency.

Once the meeting is over and all the participants have returned home safely, it's important to follow up quickly with minutes. Ensure they are comprehensive, they identify each question asked by the Agency (and who asked it) and who answered it from the team. Identify the resolutions to issues in the meeting minutes. Share the minutes with the Agency; they will be preparing their own. If there are areas of disagreement — the two parties remembered something differently or believed a different agreement was reached, this is the point to come to resolution on those issues. Although the Agency's minutes will be the official minutes, if the sponsor can get their notes to the Agency quickly enough, there will be an opportunity to resolve these conflicts. If there are areas of non-agreement that remain unresolved, I guarantee they will come back to haunt the development program.

14.1.2 Pre-IND Meetings

14.1.2.1 Objectives

Pre-IND meetings, in particular, are really designed for the sponsor and the Agency to review the design and data from the pre-clinical studies, the initial CMC information, the design of the proposed Phase I study, and the overall program and its intent. This gives the Agency insight into the scope and design of the Phase I program.

If the opportunity for the strategic planning meeting is not taken, this meeting serves as the FDA's first introduction to the product and their first impression of the sponsor.

14.1.2.2 Briefing Document

As mentioned previously, the briefing document is an important introduction to the meeting; it tells a story. The document should contain adequate background information regarding the compound, its intended therapeutic target and data supporting its rationale for use in that indication.

The document also provides an overview of the non-clinical studies that will lead to filing the IND. The "argument" developed in the document is that the data from studies conducted (as well as those not yet complete) will provide data to support an adequate risk:benefit profile for the initial clinical studies.

The document will also provide product characterization and preparation information through which the FDA can review the process and reproducibility of the methods of manufacture. Finally, the document will provide a detailed outline of the first proposed study, as well as outlining the subsequent studies.

14.2 Investigational New Drug Application (IND)

In the United States, the Investigational New Drug Application (IND) is a request for an exemption from the federal statute that prohibits an unapproved drug from being shipped in interstate commerce. Current federal law requires that a drug be the subject of an approved marketing application before it is transported or distributed across state lines. Therefore, in order to ship the drug across state lines, the sponsor must seek an exemption from that legal requirement. The IND is the means through which the sponsor technically obtains this exemption from the FDA. (See schematic of submission and review process below, Figure 14-1.)

The FDA permits the exemption after reviewing the documentation and concluding that the data provided are sufficient to indicate that it is, indeed, reasonable to proceed with certain human trials with the drug. The key aspects of the FDA's review center on the pharmacology (evidence of effect), safety (acceptable toxicity profile), Chemistry Manufacturing Controls (CMC, appropriate characterization of the product and evidence of reproducible and acceptable manufacture), the study plan (the clinical trial protocol), and the qualifications for all investigators to be involved in the study (if this is really an initial first-in-man study, the investigators are those involved at the Phase I unit in which the study will be conducted).

However, the document and its supporting materials really serve to reassure the Agency about the overall safety of this product, vis-à-vis the risks that will be introduced as testing of human subjects begins. As mentioned previously, an IND is never actually approved, it is really "allowed". In the early days of biotechnology, a number of press releases announced the "approval" of the IND. This is much less of an issue now, but it is still important to be able to convey that message to senior management.

The data provided should support the trial design, including doses used, of the initial study by providing convincing arguments that the risk:benefit ratio is clearly on the side of safety of the individuals involved.

Note that as of January 1st, 2008, sponsors submitting documents electronically are *required* to send new IND submissions in an eCTD (electronic common technical document) format. Communicate with the reviewing Division ahead of the submission to ensure that all technical issues are solved prior to submission. (The CTD format is discussed in more detail below, in the NDA section.)

14.2.1 Key Elements

There are several key elements to the IND. (See Table 14-2.)

Table 14-2. Key Elements of IND

Animal Pharmacology and Toxicology
Data suggest reasonably safe for initial human testing
Chemistry, Manufacturing, Control (CMC)
Drug Substance: Manufacturing, characterization, assays, stability
Drug Product: Manufacturing, characterization, container specifications, lot specifications and release data, assays, stability
Clinical
Investigational plan: Rationale, indications, investigative approach, subjects / patients to be enrolled, approach to risks
Investigational Brochure: Clear summary of data to date, description of expected / likely side effects
Protocol: Hypothesis, inclusion / exclusion criteria, endpoints, procedures, analysis plan, expected risks, monitoring plan and procedures

Animal pharmacology and toxicology data should support a conclusion that the drug is reasonably safe for the initial human studies (and that there is a reasonable argument for moving the drug into testing in the first place).

The primary questions related to CMC have to do with manufacturing, characterization, assays and stability of both drug substance and drug product. The point is to ensure that the patients will receive what the sponsor says they will receive at the time of dosing — from a potency and reproducibility point of view.

The Investigational Plan serves as the framework upon which the initial clinical studies are placed. Our argument to the FDA is that the study we are planning fits into an overall program that will, at the end, result in the development of this product for the targeted indication. By the way, the Agency is aware that there will be times when you will do a study that is essentially a trial balloon. "If we do this particular study and we see the following effects, then we will pursue the following development program." That approach is acceptable, assuming the safety data are acceptable. The Agency will understand the approach and will endorse studies that are somewhat beyond the overall indication for the current IND. If the results are "positive", however, the Agency would expect that a new IND would be filed to cover the new development plan.

Two other aspects which are really crucial in terms of putting the IND together are the Investigator's Brochure and the Protocol. (See the discussion about these topics in Chapter 5.) The Investigator's Brochure will provide information in a summary fashion that will be utilized by the investigators who are going to be conducting the studies so that they understand the data serving as the background to the current study. The Protocol, of course, identifies the rationale behind the particular study, as well as defining the population (more often than not for the first study, normal volunteers), the endpoints (frequent measurement of blood levels and recording of vital signs and side effects), the types of procedures to be performed, and our plan for data analysis. Monitoring procedures will also be defined and explained.

14.2.2 Review Process

Review of the IND really focuses on these two questions: Is the drug reasonably safe for administration to humans? Will the proposed protocol, as designed and written, expose subjects to unnecessary risks? The Division reviews will include a pharmacologist, a chemist and a physician (and usually a toxicologist). The process of review is formalized; consult the FDA's own internal guidelines for IND reviews (http://www.fda.gov/downloads/AboutFDA/ReportsManualsForms/StaffPoliciesandProcedures/UCM082026.pdf).

The IND is just passively approved (or "allowed"), UNLESS . . . the reviewers identify some substantial issues. The types of issues that might lead to a lack of approval might include: The subjects are exposed to an unreasonable risk, the IND does not contain enough information to assess the potential risks, the protocol is deficient in design to meet the stated objectives, the investigators are not qualified to conduct the study as outlined, and the Investigator's Brochure might be misleading or inaccurate. These issues will lead to a hold and will require specific action on the part of the sponsor.

Sometimes, the review will trigger questions that need answering prior to study initiation. That information will be conveyed to the sponsor prior to closing the 30-day window. While these types of questions might be clarification questions or might be suggestions regarding wording or types of monitoring to conduct in the protocol, they do require answers (and perhaps modifications to the protocol). It is the sponsor's job, then, to respond in a timely fashion. While generally of less concern than the issues identified in the prior paragraph, these items can likewise lead to a hold if not addressed promptly.

Recently, I had the experience of receiving a communication from the FDA at the close of their business day on the 30th day which turned out to be a Friday. The communication essentially stated that if we were willing to agree to slightly modifying the language in the protocol (it meant increased monitoring in the trial) and we replied to the Agency in a timely fashion, we could go ahead with the protocol even in advance of their formally seeing the protocol again. I had the advantage of time zones (I was on the West Coast). So, we had an internal meeting and agreed to the wording and the activities. I sent an email back to the project manager on Friday afternoon my time. By the time I had arrived in the office on Monday morning, I already had a response from the Division stating that we could proceed. Later that day, they also provided a couple of additional questions requesting clarifications, but emphatically stated that start of the protocol was not contingent on receiving answers. That is clearly the way the system should work!

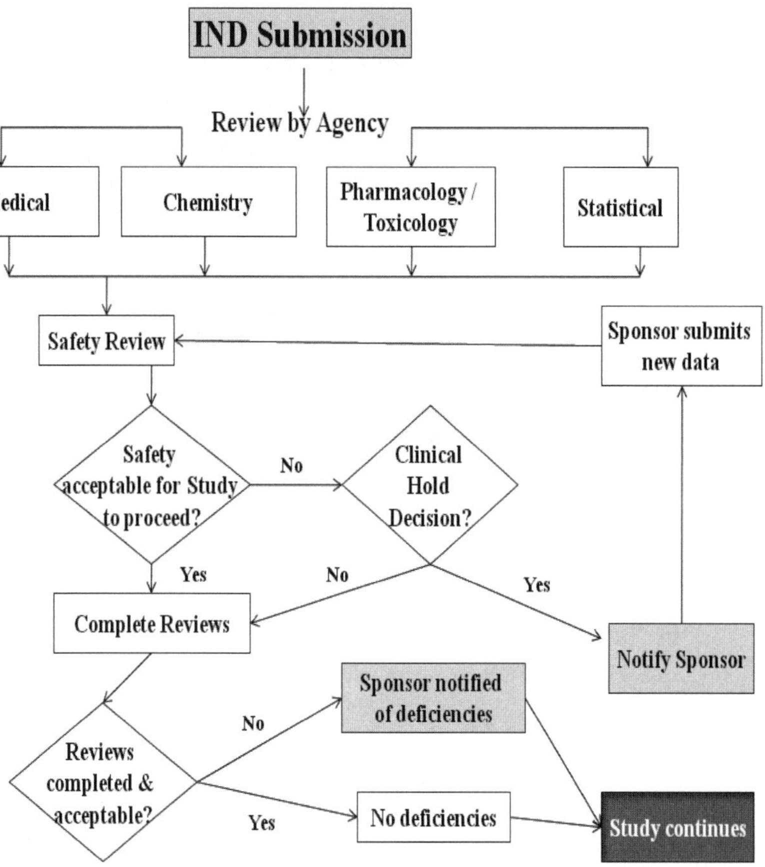

Source: U.S. Food and Drug Administration

Figure 14-1. IND Review Process

The place you don't want to be is in a **clinical hold**. You can get there because you didn't respond to questions in a timely fashion. In that case, that is your fault. On the other hand, a clinical hold could be imposed because the reviewing Division is really concerned about a safety issue that is significant enough that they are willing to go through the formal administrative hassle of imposing the hold. If that situation happens, get on the telephone as soon as possible to begin to work through the situation.

14.2.3 Types of INDs

There are two broad categories of INDs: "Commercial INDs" and "Non-Commercial INDs". *Commercial INDs* are applications that are submitted primarily by companies whose ultimate goal is to obtain marketing approval for a new product. *Non-Commercial INDs* actually represent the vast majority of INDs submitted to the FDA. The may be submitted by groups such as the National Institutes of Health (NIH) or by individual investigators. These types of INDs include "Investigator INDs", "Emergency Use INDs" and "Treatment INDs". Treatment INDs were discussed in Chapter 13.

Emergency Use IND

The need for an investigational drug may arise in an emergency situation that does not allow time for submission of an IND in the usual manner. In such cases, the FDA may authorize shipment of the drug for a specified use [21 CFR 312.36]. Such authorization is usually conditioned upon the sponsor's filing an appropriate application as soon as practicable. Prospective IRB review is required unless the conditions for exemption are met [21 CFR 56.104(c) and 56.102(d)]. Informed consent is required unless the conditions for exception are met [21 CFR 50.23].

The emergency use of an unapproved investigational drug or biologic requires an IND. If the intended subject does not meet the criteria of an existing study protocol, or if an approved study protocol does not exist, the usual procedure is to contact the manufacturer and determine if the drug or biologic can be made available for the emergency use under the company's IND.

Emergency use is defined as use of an investigational drug or biological product with a human subject in a life-threatening situation in which no standard acceptable treatment is available and in which there is insufficient time to obtain IRB approval [21 CFR 56.102(d)].

The emergency use provision in the FDA regulations [21 CFR 56.104(c)] is an *exemption* from prior review and approval by the IRB. The exemption, which may not be used unless all of the conditions described in [21 CFR 56.102(d)] exist, allows for one emergency use of a test article without prospective IRB review. FDA regulations require that any subsequent use of the investigational product at the institution have prospective IRB review and approval. FDA acknowledges, however, that it would be inappropriate to deny emergency treatment to a second individual if the only obstacle is that the IRB has not had sufficient time to convene a meeting to review the issue.

14.3 New Drug Application (NDA)

Well, this is what it is all about!! If you get this far, you and your company have some serious positive karma going.

Table 14-3. Regulatory Framework for Marketing Application Submission

Food Drug & Cosmetic Act, Section 505
21 CFR 312.3
21 CFR 314 (NDA)
21 CFR 601 (BLA)
Guidances
http://www.fda.gov/Drugs GuidanceComplianceRegulatoryInformation/ Guidances/default.htm
www.ich.org

In a legal sense, the New Drug Approval Application (NDA) represents the sponsor's request for approval for sale and marketing in the United States of the new drug for its proposed use. (See Table 14-3, Regulatory Framework.) The request is based on the documentation and analysis presented in the application and its supporting information.

14.3.1 Common Technical Document

The world has turned a few times since I entered the industry and we now submit NDAs in a CTD format. Below we will review the CTD format primarily for the purpose of understanding how the elements in the development program actually fit together to create a coherent and compelling package. More detail is provided in Chapter 10 in the discussion of data flow from patient to submission.

To reiterate, the CTD is divided into 5 modules. (See Figure 14-2.)

- Module 1 represents region-specific (e.g., US / FDA) administrative information that is specified by the regulatory agency to which you are applying;
- Module 2 represents of a summary of all the documentation "below" in the pyramid;
- Module 3 represents Quality (chemistry and pharmaceutical development)
- Module 4 represents Safety (non-clinical data); and,
- Module 5 represents Efficacy (clinical reports, including human safety).

Graphically, Figure 14-2 represents the organization of the CTD as a triangle or pyramid. The foundation or floor of the pyramid, Modules 3, 4, and 5, is comprised of the individual reports and the core data. Above these sections are placed the summaries corresponding to the sections below. So, Quality is summarized in Module 2.3; the non-clinical summaries are provided in Module 2.6 and further summarized in 2.4; the clinical summary is located in Module 2.7 and the clinical overview in 2.5. The overall introduction (executive summary) is located in Module 2.2 and is preceded by the table of contents in 2.1. As noted above, the information in Module 1 is driven by the regional authority to which the application is actually submitted.

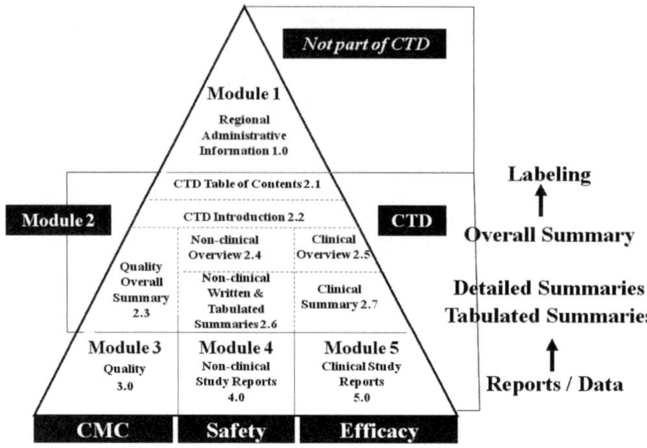

Figure 14-2. Organization of the ICH CTD.

14.3.1.1 Module 3

Module 3 is the Quality section; this corresponds to CMC (chemistry, manufacturing, controls). Module 3 presents, in the appropriate sections, all of the essential data relating to the drug substance and to the drug product. Section 3.2.S presents information about the manufacturing, characterization, assays, and stability of the drug substance. Section 3.2.P focuses on the drug product and provides information regarding manufacturing, characterization, container specifications, lot specifications and release data, assays, and stability. The Appendices to Section 3.2 deal with facilities and any adventitious agents or novel excipients that are important to the manufacturing or preparation process.

14.3.1.2 Module 4

Module 4 is called Safety, but contains the non-clinical information. Hence, it focuses on all the data and study reports pertaining to pharmacology, pharmacokinetics, and toxicology. Sections 4.2.1 (Pharmacology), 4.2.2 (Pharmacokinetics), and 4.2.3 (Toxicology) contain individual study reports and Section 4.3 contains appropriate literature references for these topics.

14.3.1.3 Module 5

Module 5 is called Efficacy, but contains the detailed information pertaining to clinical studies and, despite its designation, includes a discussion of safety derived from the clinical trials. Section 5.2 lists all clinical studies. Section 5.3 contains all clinical study reports arranged in the following order:

- 5.3.1 Biopharmaceutic studies
- 5.3.2 Pharmacokinetics using human biomaterials
- 5.3.3 Human pharmacokinetics
- 5.3.4 Human pharmacodynamic studies
- 5.3.5 Efficacy and safety reports for indication (controlled, uncontrolled, meta-analyses, and other study reports)
- 5.3.6 Reports of post-marketing experience
- 5.3.7 Case Report Forms and individual patient listings

Section 5.4 contains literature references applicable to the clinical studies.

14.3.1.4 Module 2

Moving up to Module 2 — this module represents progressively higher level summaries of the information contained in Modules 3, 4 and 5. It is here that the arguments are constructed that "sell" the benefit: risk characteristics of the drug in the patient population for which we are seeking marketing approval.

The sections are ordered in the following manner:

- 2.1 CTD Table of Contents
- 2.2 CTD Introduction
- 2.3 Quality: Overall Summary
- 2.4 Non-Clinical Overview
- 2.5 Clinical Overview
- 2.6 Non-Clinical Written and Tabulated Summaries
- 2.7 Clinical Summary

Precise formatting discussions are way beyond the scope of this book. It is best to have in-house or contract experts who can assist the development of the documents from the ground up at the earliest stages of development. Doing that preparatory work initially will, in my opinion, greatly facilitate your development (including applications for clinical testing in various countries) and ultimate approval.

14.3.1.5 Module 1

As noted previously, Module 1 represents the administrative information required by the Agency to which the application is sent. For applications submitted in the US, the FDA has specified the information to a certain extent, and the main aspects are identified below.

- Marketing Application Form (US = Form 356h)
- Tracking of regulatory interactions and fulfillment of regulatory requirements / requests
- Proposed product labeling
  - Use Target Product Profile as guide
  - Target Product Profile Guidance (http://www.fda.gov/downloads/Drugs/GuidanceComplianceRegulatoryInformation/Guidances/ucm080593.pdf)
- Comprehensive Table of Contents for All Modules

14.3.2 Regulatory Review

Any marketing application has as its goal the marketing and sale of a particular product. The information contained in the application should permit the FDA to decide whether the drug is safe (do the benefits outweigh the risks?), whether it is effective for the proposed uses, whether the proposed labeling is adequate and appropriate (are the target, indication, usage and data accurately portrayed?), and whether the product can be manufactured reliably (are methods adequate to preserve the drug's identity, strength, quality and purity?). If the sponsor has done its job, the review should be straightforward.

It is really crucial that the NDA be intelligible. I have actually seen NDAs for which it was clear that no single human read the entire document. Sentences from the prior paragraph were repeated verbatim, then sections were repeated verbatim within the same portion of the document (the wonders of cutting and pasting). References to other sections would lead to a different place or, worse, to nothing. Those responsible for assembling the document need to make sure it's written in English that people understand.

Likewise, I believe it is crucial that somebody read the document all the way through from beginning to end to make sure it makes sense. It's also important to have somebody literally go through just checking each of the references to make sure that each one refers to the correct location. Newer formatting with hyperlinks makes checking much easier. On the other hand, if the hyperlinks don't work or the reference is wrong, the problems are glaring.

Have some sympathy for the reviewers. Make their lives a bit easier.

14.3.2.1 Review Process

NDAs are reviewed in the following way. The project manager (we used to call them consumer safety officers, CSOs) is assigned. This tends to be the same person involved with the sponsor along the way, unless personnel have changed.

The first review process assesses the "suitability" of the NDA, which really translates to something called "fileability". The FDA, at that point, makes a designation and says this document is "fileable". Now the fileable designation, of course, does not mean that they are going to approve it. It just means that the required elements are in place; they have paged through and reached the conclusion that there are sufficient data from which an assessment can be made. Again, this is not necessarily a positive conclusion. That date becomes the filable date, which is typically somewhere in the vicinity of two, three, or four weeks after the date that the NDA was delivered to the Agency. Companies refer to filing dates as the date stamped on the receipt at the Agency when delivered. The FDA's filing date refers to this "suitability" review. By the way, if the Agency doesn't think the submission is suitable, it will send a letter within 60 days — frequently sooner — and will keep 25% of the submission fee.

Though FDA reviewers are involved with a drug's development throughout the IND stage, the official review time is the length of time it takes to review a New Drug Application and issue an action letter, an official statement informing a drug sponsor of the Agency's decision, following submission of the NDA.

Once a New Drug Application is filed, an FDA review team — physicians, chemists, statisticians, microbiologists, pharmacologists and other experts — evaluates whether the studies the sponsor submitted show that the drug is safe and effective for its proposed use. Since no drug is absolutely safe, "safe" in this sense means that the benefits of the drug appear to outweigh the risks.

In the primary review process, reviewers attempt to confirm and validate the sponsor's conclusion that a drug is safe and effective for its proposed use. The review is likely to involve a reanalysis or an extension of the analyses presented in the NDA. For example, the medical reviewer may seek to reanalyze a drug's effectiveness in a particular patient subpopulation not analyzed in the original submission. Similarly, the reviewer may disagree with the sponsor's assessment of evaluable patients and seek to retest effectiveness claims based on the patient populations defined by the reviewer.

Review team members communicate extensively with each other. If a medical reviewer's reanalysis of clinical data produces results different from those of the sponsor, the reviewer will forward this information to the statistical reviewer with a request for a statistical reanalysis of the data. Likewise, the pharmacology reviewer may work with the statistical reviewer in evaluating the statistical significance of potential side effects in long-term animal studies.

When the technical reviews are complete, each reviewer prepares a written evaluation containing conclusions and recommendations about the

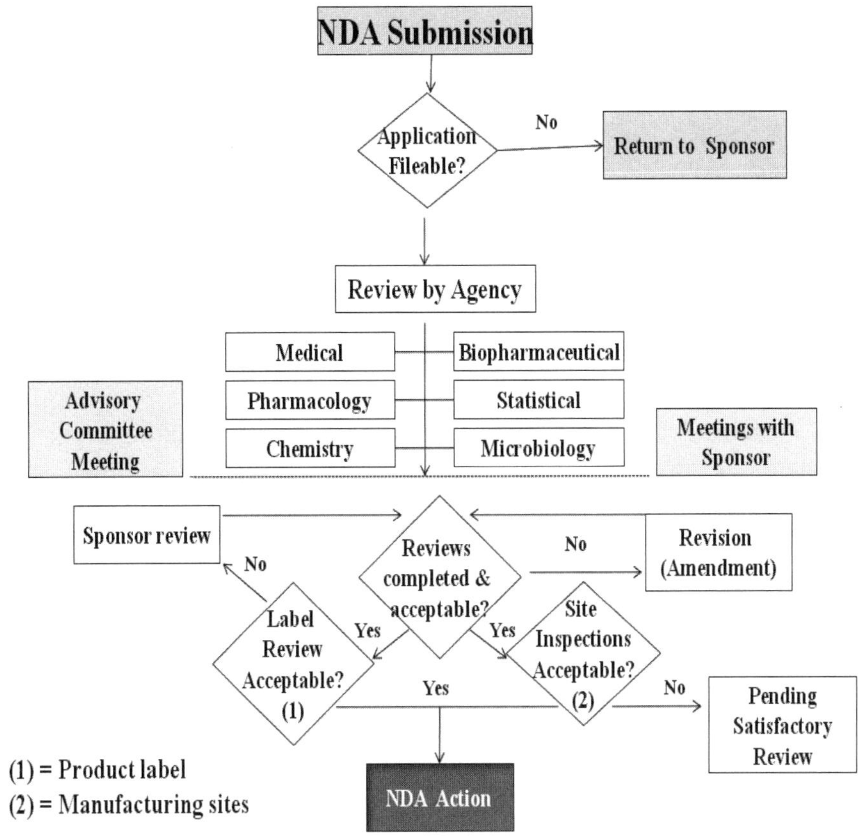

Source: U.S. Food and Drug Administration

Figure 14-3. NDA Submission and Review Process

application. These evaluations are then considered by team leaders, division directors, and office directors, depending on the type of application.

If the FDA decides that the benefits of a drug outweigh the risks, the drug will receive approval and can be marketed in the United States. But if there are problems with an NDA, the FDA may decide that a drug is "approvable" or "not approvable".

A designation of "**approvable**" means that the drug can probably be approved, provided that some issues are resolved first. Typically, this requires that the sponsor and the FDA reach a final agreement about specific wording in the product's label. It could also involve more difficult issues, such as the adequacy of information on how people respond to various dosages of the drug.

A designation of "**not approvable**" describes deficiencies significant enough that it is not clear that approval can be obtained in the future, at least not without substantial additional data.

Common problems include unexpected safety issues that crop up or failure to demonstrate a drug's effectiveness. A sponsor may need to conduct additional studies — perhaps studies of more people, different types of people, or for a longer period of time.

Manufacturing issues are also among the reasons that approval may be delayed or denied. Drugs must be manufactured in accordance with standards called Good Manufacturing Practices, and the FDA inspects manufacturing facilities before a drug can be approved. If a facility isn't ready for inspection, approval can be delayed. Any manufacturing deficiencies found would need to be corrected before approval.

The FDA outlines the justification for its decision in an action letter to the drug sponsor. When the action is either approvable or not approvable, the FDA gives the sponsor a chance to meet with Agency officials

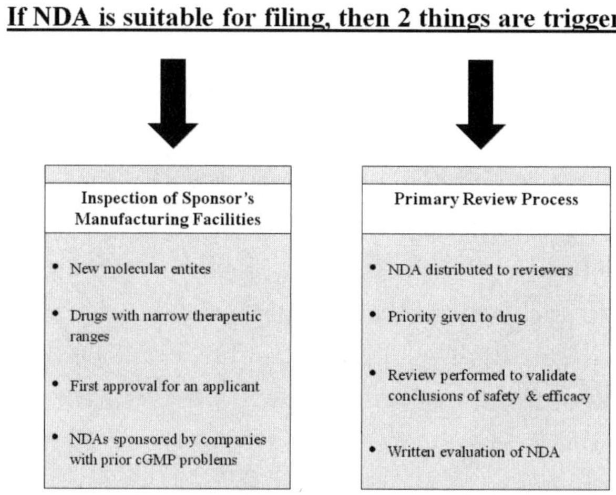

Figure 14-4. Activities Once NDA is "Filed"

to discuss the deficiencies. At that point, the sponsor can choose to ask for a hearing or correct any deficiencies and submit new information.

Once the NDA is reviewed or the review process has begun, then a process begins which 1) triggers inspections of the sponsor's manufacturing facilities, and 2) initiates the overall review process. The details of those processes are more complicated than we have time to address here. The key message is that once the NDA is deemed filable, then those things begin to occur.

Inspections

The Food and Drug Administration relies on data that sponsors submit to decide whether a drug should be approved. To protect the rights and welfare of people in clinical trials, and to verify the quality and integrity of data submitted, the FDA's Division of Scientific Investigations (DSI) also conducts inspections of clinical investigators' study sites. DSI will, in addition, review the records of institutional review boards to be sure they are fulfilling their role in patient protection. DSI seeks to determine such things as whether the study was conducted according to the investigational plan, whether all adverse events were recorded, and whether the subjects met the inclusion / exclusion criteria outlined in the study protocol.

At the conclusion of each inspection, FDA investigators prepare a report summarizing any deficiencies. In cases where they observe numerous or serious deviations, such as falsification of data, DSI classifies the inspection as "official action indicated" and sends a warning letter or Notice of Initiation of Disqualification Proceedings and Opportunity to Explain (NIDPOE) to the clinical investigator, specifying the deviations that were found.

The NIDPOE begins an administrative process to determine whether the clinical investigator should remain eligible to receive investigational products and conduct clinical studies. CDER conducts about 300–400 clinical investigator inspections annually. About three percent are classified in this "official action indicated" category.

14.3.2.2 FDA Classification System

Once the Agency has initially seen the drug (at the time of the IND submission), they tend to apply a ratings system to the product. The product is classified by rating its chemical novelty and by assessing its therapeutic potential (priority, standard, special situation).

Occasionally, they will assign a special situation status for a particular product, and that assessment tends to hold throughout the course of the development. It is important that, if the sponsor thinks there is something particularly unique about the product — either in terms of its overall profile, its overall efficacy or its overall safety in the designated patient population — the sponsor identifies that early in discussions with the Agency to improve the chances of achieving that priority designation.

NDAs are classified by the FDA with a code that reflects both the type of drug being submitted and its intended uses. The numbers 1 through 7 are used to describe the type of drug:

1. New Molecular Entity
2. New Salt of a Previously-Approved Drug (not a new molecular entity)
3. New Formulation of Previously-Approved Drug (not a new salt OR a new molecular entity)
4. New Combination of Two or More Drugs
5. Already-Marketed Drug Product — Duplication (i.e., new manufacturer)
6. New Indication (claim) for Already-Marketed Drug (includes switch in marketing status from prescription to OTC)
7. Already-Marketed Drug Product — No Previously Approved NDA

The following letter codes describe the review priority of the drug:

S. Standard review for drugs similar to currently-available drugs

P. Priority review for drugs that represent significant advances over existing treatments

14.4 FDA Advisory Committee Meetings

The FDA uses a group of committees called advisory committees to provide review and advice on a number of topics. One situation of specific importance to a company occurs when their drug is presented to the advisory committee prior to final FDA review.

These committees are typically therapeutic area specific. For example, there would be a cardiovascular advisory committee, a dermatology advisory committee, an oncology advisory committee, etc.

The FDA uses these committees to obtain outside advice on technical and medical issues related to regulated products. It also uses the advisory committee to discuss issues that the FDA believes are important to public health. For example, a topic on an agenda of an advisory committee might be the use of a class of drugs, such as cox-2 inhibitors. The advisory committee, in this case, is providing their thoughts and their advice about a larger question than reviewing a specific therapeutic agent. Use by the FDA of advisory committee meetings in these settings encompasses both scientific and, in reality, political issues. In certain instances, I would suggest the Agency is essentially spreading the risk in a making a certain decision — more input, increase the list of blameable folks.

Advisory committee meetings will also always happen when this is the first entry into a pharmacological class — either a new mechanism of action or new chemical structure that the Agency hasn't seen before. If the Agency believes the product may have major therapeutic impact in an important disease, or even range of diseases, they will bring it to an advisory committee. Generally, first in class OTC switches will also come before the committee.

Thus, advisory committees, and the process involved with getting there, are important elements for consideration in putting together the development program. If development is successful, it is likely the product will be presented by the FDA to an advisory committee. Thus, it is good to have an idea of what it entails.

Membership on the advisory committee depends on the charter of the particular committee and the technical expertise required. In general, members tend to be physicians — mostly because the committees are the next-to-last step before the product gets final approval and moves into the marketplace. It largely has to do with the clinical use of a particular product. On the other hand, there are consumers who are part of certain committees as well as industry representatives. And, they do not necessarily have to be physicians.

It's also important that the advisory committee group have no conflict of interest. We have talked previously about potentially using advisory committee members as consultants, and we suggested that a good group to look at would be those people in the last year or two of their terms at the point in time that the sponsor begins to work with them, because they will be off the committee by the point in time that the product comes up for review. Therefore, that conflict of interest would be removed.

Terms are generally four years. They may be renewable, particularly when a person is a member of the committee and then subsequently becomes chairman of the committee. The FDA provides the executive secretary for the committee, who is a non-voting member.

14.4.1 Prior to the Meeting

Prior to the meeting, the FDA will prepare an analysis of the marketing application, prepare a summary, and provide that to the committee. The sponsor generally also summarizes their analysis and opinions and submits a briefing package for the committee to consider.

One item that is important for sponsors to understand it that almost all of the information presented in the context of the committee meeting is publicly available at the time of the meeting; hence, the package the sponsor submits should only include information the sponsor is willing to publicly disclose. One possibility is to identify certain material (should represent a small proportion of the information submitted) that is truly proprietary (e.g., product design features for a device) that would be exempt for the Freedom of Information Act.

Once there is a date, a time, a location, and a spot on the agenda, what is next?

Practice as though life itself depends upon the outcome!! If the sponsor is a small company and this is the first — perhaps only — product, it just may represent survival. So, as for other meetings, practice, practice, practice and practice.

We discussed some of these suggestions previously when we talked about development meetings with the reviewing division. The logic is the same — the sponsor shouldn't be asked a question for which it hasn't prepared. If preparation for other meetings with the FDA seemed like a big deal, preparing for an Advisory Committee meeting is a *really* big deal. Get whatever help you need! Keep in mind, this is a very public forum, the presentations must be very smooth, well-organized, and professional — although not too glossy.

The future of the product (and sometimes the future of the company) rides on the outcome of the meeting. The absolute best people possible need to be presenting. The presenters might be people from the company, investigators or consultants (yes, I am available for hire!), but they need to be very, very good.

Many times the presentation may include multiple presenters, e.g., the company person (or consultant) does the introduction and summary and perhaps much of the information in the middle. However, an engineer might present data related to failure testing of a device, or an investigator might present data from one or more trials (remember, this could be the person who was previously on the advisory committee who was hired as a consultant a long time ago). Whatever transitions occur during the presentation must be seamless — again, the key is practice.

In addition, the team should include someone who will direct questions from the committee to the appropriate person for answers and will include someone identified to take notes. The roles and activities should all be sorted out in advance.

By the way, always take back up equipment — cost is irrelevant at this point. Take a back up computer, projector, memory sticks, CDs, DVDs, microphones, pointers, etc.

It is also important — prior to the meeting — to learn as much as possible about the committee members, especially the chairperson. This isn't a covert subterfuge exercise. It is more from the standpoint of understanding the audience. Identify their particular expertise and any particular biases that might need to be considered during the presentation. An example of the latter might be an editorial the person had written that, in some way, is relevant to the story being presented. Many of these people will be in the academic community and will have published information. Study the agenda. The FDA puts this together in a very considered way. It is a publicly available piece of information. Understand exactly how much time is available for the presentation. Remember our guidance about the number of slides: Use approximately three minutes per slide.

Understand what other products are being considered at the same time. Frequently, there is a reason why multiple products are considered at the same time — the products may raise the same issues, be in a particular therapeutic area, or address a particular therapeutic approach, etc. On the other hand, sometimes they are somewhat random and they are simply dictated by the chronology, the applications and the review.

Try to understand how this committee functions. For example, if the sponsor is developing a cardiovascular product, it makes sense — early in development — to identify the panel members and the panel's meeting schedule. It would make sense to attend some of the meetings and to review the minutes (they are posted on the FDA website); this provides an insight into the committee operation, the key questions members ask, and issues that seem important to them.

14.4.2 At the Meeting

Remember, the Advisory Committee Meeting is an open public hearing and discussion.

The FDA and the sponsor will do presentations. The access is unrestricted. That is, all competitors may attend. There is no public participation — meaning a competitor can't ask a particularly noisome question — for a product review. On the other hand, all attendees fully understand the discussion that's going on.

There may be a closed presentation of data, which will include only the FDA, the committee and the sponsor. If the sponsor needs to discuss some data in private, that request must be identified to the Agency at the time the sponsor first becomes aware of their appointment with the Committee. The committee deliberations themselves will happen in a very closed environment, which will include only the FDA and the committee.

So, what happens as a result of this particular meeting? The Committee will make its decisions and recommendations. The decisions and recommendations will be very specific and address precisely the questions presented to the panel by the FDA.

The FDA will typically provide its opinion to many of the issues being discussed. The issues, as we talked about previously, could range all the way from the product for this particular indication to a larger question, which may have impact for the particular product. But, in any case, the FDA will opine as to what their position is. Then, the committee will provide its recommendation.

There are certainly well publicized instances in which the Committee's recommendations are not the same as the FDA's. Accordingly, it is important to understand that the committee recommendations are not binding on the FDA. However, the FDA never ever ignores or disregards the recommendation. It will regard them, although the FDA may make a contrary decision.

14.5 Post Approval

14.5.1 On-Going Regulatory Operations

Section 11.3.2 covers many of the activities required to "service" the IND on an on-going basis.

Post-approval, there are still requirements for monitoring safety. The goal of risk assessment post-approval is to continually re-assess the benefit:risk profile in the face of data generated in the context of broader use. One of the challenges is gathering the appropriate data to put product use into context. Safety data derive from a variety of sources, ranging from spontaneous reporting of adverse events to larger observational studies, such as those required as conditions of the product's approval.

Safety evaluation in the context of broad market use is a very complex process requiring diligent collection, processing, and analysis of relevant data enabling appropriate frequent re-assessment of the benefit:risk ratio. As long as it remains heavily on the benefit side of the equation, the product contributes to the well-being of patients. If, for some reason, the scales tip to the risk side, then changes in the label might need to be made, or worse, the product might need to be removed from the market.

On-going iterative risk assessment is a crucially important activity that continues, essentially, throughout the lifetime of the drug.

14.5.2 Advertising and Promotion

From the point in time there is some inkling the product will be successful, the sponsor needs to consider advertising and promotion — this is true even if the sponsor is planning to partner the asset. Competition (present and future) and third-party reimbursement groups must be kept in mind. Ultimately, the sponsor (or whoever is responsible for marketing) needs to give consumers (patients, physicians, reimbursement agencies) a reason to choose this product versus another product or treatment protocol.

Once a product is approved for marketing, advertisements related to that product must contain a brief summary. The brief summary is defined as the list of side effects, contraindications, warnings, precautions and indications. If the product label contains warnings (e.g., black box warnings), that information must be clearly displayed in any type of advertisement released into the public domain.

Claims, direct or implied, cannot be made beyond what is in the product label. Anything not contained in that product label or that package insert is considered off-label promotion and is seriously frowned upon by the FDA and the Federal Trade Commission.

In terms of advertising materials, there is no FDA pre-clearance of the material prior to the time of approval (that is, FDA approval for marketing), except in cases of accelerated approval. However, each time product advertising is assembled this must be forwarded to the FDA for review (attached to FDA Form 2235 and including a copy of the package insert and the promotional material to be used). In today's marketing environment, that material could be videotape, CD, DVD, print media, podcasts, and banner ads used on the internet. All of those vehicles of dissemination are covered under this advertising and promotion regulation.

There is an interesting split present in the device world. The Federal Trade Commission governs what are called non-restricted devices, while what are called restricted devices are governed by the Food and Drug Administration. Perhaps medication for schizophrenia might be helpful.

Compliance requires that all advertising must adhere to the data available. If data aren't available to support the statements, the statements can't be made!

All of this information, at one level, is intuitively obvious; however, it doesn't keep people from trying to edge past the regulations. I believe it's important to mention this within the context of doing clinical trials, particularly, as we consider small companies. Because there is so much pressure for all small companies to have positive data, to have information that enables them to go to their investors and ask for more money, to move a product closer to the marketplace, to achieve a substantial financial reward, there is tremendous pressure to stretch the truth. Just say, "No."

I have a friend, a person with whom I used to work. He ultimately became CEO of a company. A couple of years ago, he and his company received a warning letter from the Food and Drug Administration, because, three layers down from the home page, there were statements suggesting superiority to competitive products. Data supporting those statements were not available. It is reasonable to surmise that the information was highlighted by the competition. But, it doesn't matter. The FDA reviewed the statements, looked at the data they had, contacted the company and suggested, ever so politely, that the company needed to close down its website until the problem was fixed (and reviewed by the Agency before coming back on line). By the way, there was a hefty fine.

14.5.3 Reimbursement

Starting with the concept of "Begin with the End in Mind", the sponsor needs to understand the reimbursement requirements for each market it plans to enter. The US is different than Canada, is different than the EU, is different than Japan, is different than other markets. While the CTD process facilitates the application process worldwide, approvals are for individual markets and within those markets there are separate reimbursement processes. That is true in the US and it is likewise true in the EU. Sponsors absolutely must plan for reimbursement in the development program.

Fortunately, what is good for the reimbursement process is generally good for advertising and marketing.

14.6 Summary

By way of summary, allow me to restate some clear biases of mine, expressed initially in Chapter 13.

Regulatory strategy is a competitive weapon. Regulations around the world are not all black and white. There is a lot of grey in those regulations, and the grey enables the development team to think uniquely and creatively about the product development. Ultimately, this question needs to present itself and be answered: "How do I get the maximum commercial benefit from my product ethically, legitimately, honestly?" Exploit all the regulations you can in a responsible way.

Part IV

Changing the Plan: Trust Me, The Future Will Be Different

Chapter 15

Where Do We Go from Here?

The common thread of the topics in this session is that all will have something important to do with the conduct of clinical trials in the future.

15.1 Technology

First, we will briefly touch on the implications of technology for the future and then discuss four topics within that context: personalized medicine, FDA's Critical Path Initiative, adaptive clinical trial design, and bioinformatics and electronic data. These topics will, in my opinion, significantly affect future processes of product development, some more pervasively than others.

Our goal is to get our products to market. But in considering new technology, the issues of regulatory compliance, security, confidentiality, something I call dependency — or actually inter-dependency — time and cost will be important considerations.

Where do we find ourselves today? First, there are a lot of *issues in clinical practice.*

There is intense discussion at present about overhauling the health care delivery system in the United States. There is concern that many people die each year due to medical errors. Many physicians rely entirely or largely on hand-written medical records.

When I was still a real doctor, a serious problem was the inability to decipher physicians' handwriting (other physicians' handwriting; not mine, of course!). At the time, a particular problem was the confusion between Isordil™ and Inderal™ — both cardiovascular medications. The dosages could be similar, so confusion was even more likely. For the patient, though, the differences were not subtle!

Fast forward several years and I was attempting to build a business that depended upon the presence of electronic medical records in physicians' practices. I can attest to the small numbers of physicians using electronic records. You also might remember that President George W. Bush made electronic health care records an important component of his health care policy.

As detailed in Chapter 1, there are huge *bottlenecks in development.*

Technology is changing rapidly. In the industry, there is a fundamental and profound need for high-quality, secure and unimpeachable data. The US FDA has promulgated a rule that has been codified — 21 CFR Part 11 — dealing with electronic records and electronic signatures. The FDA — and many IT experts — remains less than fully convinced about the reliability and security of electronic signatures. Nevertheless, the ICH harmonization process has assumed that all files and data will be electronic. Nevertheless, considering the standard review, revision, and issuance process for guidance documents, guidance documents devoted primarily to technical issues tend to lag behind current developments.

So, *security* is a concern. That being said, industry generally overestimates the value of the individual pieces of data — particularly clinical data. For anyone who has ever pored through tables and listings of clinical data, it is obvious that even large chunks of the data probably wouldn't make much sense outside the context of the situation in which they were collected.

Indeed, more than three-quarters of the damage to databases in the pharmaceutical industry comes from within the organization — some of it is wholly unintentional. On the other hand, the intentional damage is largely at the hands of disgruntled employees. While there are a lot of ways to protect data, they still depend on the humans involved.

Have you ever sat at your local friendly Starbucks or in the comfort of your own home, opened your computer and turned on your wireless connection to see multiple other networks in your vicinity? How many of you at Starbucks, at home, or in the airport are transmitting "sensitive" data on a regular basis?

Confidentiality is the next issue to rear its ugly head. Just for the sake of being clear about the definitions — privacy refers to personal disclosures and confidentiality to sharing private information. Earlier in the book (Chapter 7), we discussed HIPAA. I said then that HIPAA changed everything when it came to clinical trial conduct - I stand by that statement.

It is perhaps intuitive, but it bears stating — we are all in this together. We have significant *inter-dependency*. Patients, hospitals, practitioners, pharmaceutical and device companies, investigative sites and regulatory authorities are inescapably intertwined. What affects — good or bad — any of them affects the others.

Currently, there are multiple data formats, each of which seems to make sense in its individual context — hospital, practice, company — but, in combination, they lead to chaos.

The internet has changed our world in so many ways! Of course, medicine and drug development have likewise been impacted. Patients now can learn a great deal about their own diseases, current therapy, experimental approaches, and can play a significant role in their own care. Considering the general state of medical care these days, that is probably a really good idea! The difficulty, however, is that not all of the information on the internet is reliable and is, sometimes, just plain incorrect. Given that the majority of usage of the internet by the elderly concerns health care, this can be a problem.

Most of us are worried about enrolling our 50-patient or 150-patient or perhaps 500-patient trial, but on an annual basis, there are more than 15,000 clinical trials being conducted worldwide involving nearly a half million people at any given time. No wonder any given sponsor has trouble getting its trial enrolled!

15.2 Topics to Watch

From here, let's consider four topics that I think will be important in the near future: personalized medicine, the Critical Path Initiative, adaptive clinical trials, and finally, bioinformatics and electronic data.

15.2.1 Personalized Medicine

Personalized Medicine: Is it hope or hype? Curiosity or opportunity? The quick answer is, probably a bit of each.

15.2.1.1 Definition

Personalized medicine sometimes goes by the names pharmacogenomics or pharmacogenetics. Regardless of the term, what is it?

Here is one definition: "The management of a patient's disease or disposition by using molecular knowledge (genomics, DNA modification, proteomics, metabolomics) to achieve the best possible medical outcome for that individual. It includes improved prevention, diagnosis, treatment, and monitoring across both medical research and clinical care." (McKinsey & Company) It doesn't quite roll off the tongue!

To me, personalized medicine comes down to providing the right medicine in the right dose to the right patient at the right time.

15.2.1.2 Development Context

Who really cares? To get to the answer to that question, we need to determine what problems we are trying to solve.

If one considers different classes of drugs, there are inconsistent responses (i.e., differential efficacy) across populations to various medicines. Some examples are in Table 15-1. Perhaps one-third of patients with hypertension respond well to angiotensin converting enzyme (ACE) inhibitors. Maybe one-quarter of patients with congestive heart failure respond to beta blockers.

Statins have been the wonder drugs of the last two decades, yet a significant fraction of people have little response in LDL levels and, more

Table 15-1. Differential Efficacy

The response rates to medicines are inconsistent.	
ACE inhibitors (hypertension)	10–30%
Beta blockers (heart failure)	15–25%
Statins (cholesterol)	30–70%
Beta-2 agonists (asthma)	40–70%

Source: Personalized Medicine Coalition

importantly, little change in cardiovascular risk. In asthma, nearly 40% of patients don't adequately respond to beta-2 agonists.

There is nothing in the rule book that says that everyone has to respond. However, there are concomitant issues of safety. Adverse drug reactions occur frequently; indeed probably more than two million cases are reported each year. More than one-quarter of hospitalized patients have drug-related adverse effects. The FDA suspects that in excess of 100,000 people in the US die annually from drug-related adverse events. The costs to our health care system of these events are staggering. (Source: F. Frueh, FDA)

In addition, there are a lot of problems we have not seriously addressed. There are more than 6,000 orphan diseases identified. It is likely that potential drug targets represent some 3.5% of the human genome (perhaps some 1,000 genes). Yet, 50% of the drugs on the market today target only four key gene families. Even new molecular entities are largely directed against old targets.

We examined the overall success rate for development products in the Chapter 1. Even in the clinical trial stage, the success rate is abysmal. Huge trials are very common, but productivity is not good. (See Table 15-2.)

15.2.1.3 Benefits of Personalized Medicine

Within that context, what might the benefits of personalized medicine be?

Table 15-2. Clinical Trials Are Too Big and Too Costly

Failure rates of drugs in clinical development are high. (drugs entering development in the mid-1990s)	
Phase 1	30–40%
Phase 2	60–70%
Phase 3	45–55%
Submission	10–20%
→ Cumulative Failure Rate > 95%	
• 10–40, 000 patient trials are common	
• ↑ R&D Expenditures, but ↓ NCE approvals	

Source: Wood, NEJM, 2006; PhRMA; FDA

It might be possible to detect a disease process at an earlier stage, rendering it more susceptible to treatment. Perhaps it would enable a better selection of more optimal therapy. It might enable us to significantly reduce adverse drug reactions and, in combination with better efficacy, might lead to better patient compliance.

Earlier in the process, it might enable improved drug discovery target selection and reduce the time, cost, and failure rate through all of development, particularly in clinical development.

From more of a big-picture perspective, personalized medicine might also shift the focus of medicine from a reactive treatment process to a more proactive preventative care model, and would potentially reduce the overall cost of delivered health care.

There are actually a series of successes that supporters of personalized medicine can point to. (See Table 15-3.)

Probably the most notable one was the development of Herceptin® and the HER2 diagnostic test based on the discovery of the HER receptor and the discovery that HER2+ breast cancer was a separate entity. Gleevec® and the bcr-abl diagnostic represented another success enabling more focused and more successful treatment of chronic myelogenous leukemia and stromal tumors of the gastrointestinal tract.

Table 15-3. Personalized Medicine Successes (examples)

Test	Application
Herceptin® and the HER2 diagnostic test	HER2+ breast cancer
Gleevec® and the bcr-abl diagnostic	CML and GI stromal tumors
Amplichip®	Cytochrome P450 (CYP2D6 & ↓ tamoxifen response rate)
UGT1A1 Assay	Irinotecan and colorectal cancer
Thiopurine methyltransferase test	ALL, monitor toxicity
BRCA 1 or BRCA 2	Increased risk of breast cancer
p16	Susceptibility to melanoma
OncotypeDX24®	Support treatment and monitoring decisions for cancer

The Amplichip® facilitated testing of the cytochrome p450 enzyme system (enzymes present in the liver that contribute to the metabolism of a large number of drugs). By facilitating the testing on a clinical level, we discovered that some patients with breast cancer and decreased responsiveness to tamoxifen had different levels of CYP2D6 — one of those p450 enzymes — than people with "normal" responsiveness.

Another assay, this one for UGT1A1 — another metabolic enzyme in the liver — enabled a better understanding of differential responsiveness to irinotecan in patients with colorectal cancer.

While we are early in the personalized medicine story, there are still some clear issues to address as we move down this pathway. First is the sheer enormity of the task in front of us. Research budgets in companies and academic institutions are largely inadequate to fully pursue additional biomarker and translational research. Despite the success stories described in Table 15-1, the truth is that we have inadequate economic impact information confirming the benefit of the various personalized medicine approaches.

Woefully lacking is physician education and other health care provider education regarding the personalized medicine area. A substantial fraction of physicians, at least, were not exposed to the key genetic concepts in medical school and thus are unprepared to fully understand the subtleties of the area.

Next — and a major concern — is the lack of a sound regulatory framework within which to review and approve biomarker-disease combination products and diagnostics. In addition, the lack of privacy guidelines and legislation to protect personal information and limit insurance and employer actions related to the information obtained is a problem.

The reimbursement agencies have generally been slow to respond to developments in this area and seem, in some cases at least, to have limited understanding of the technology.

Finally, and a major issue, is that there is an inadequate information technology framework to acquire, aggregate, integrate, analyze and understand data related to personalized medicine.

So, if there is any hope of getting past the hype and the curiosity to capitalize on the opportunity, we need to address the challenges (among others) outlined in Table 15-4.

Funding needs to be provided for well-designed studies in pharmacogenetics. We also need to find a way to connect information from earliest discovery through post-approval drug surveillance. This one area I know well — several years ago, I tried to build a company that attempted to do this. We were then, and probably would still be today, too early.

A corollary to the last comment is the need for electronic medical records. I believe strongly that electronic medical record implementation needs to happen, first for a safety reason and then, as a means of deriving full benefit from personalized medicine and other IT-dependent activities.

Table 15-4. Challenges for Personalized Medicine

- Provide funding for well-designed studies in pharmacogenetics
- Connect information from earliest discovery through post-approval drug surveillance
- Expand (and ultimately make universal) use of electronic medical records
- Establish the appropriate regulatory framework on a national and international level for evaluation and approval of pharmacogenetic tests
- Harmonize regulatory approaches toward transmission, storage and processing of large amounts of complex data
- Harmonize guidelines and regulations for conduct of genetic research across international borders
- Expand education in understanding and interpreting broad pharmacogenetic information
- Fund integration of IT, data management, and analytics for better understanding of information
- Demonstrate economic value of these approaches to foster the near-term investment

Importantly, an appropriate regulatory framework needs to be established on both a national and an international level for the evaluation and approval of pharmacogenetic tests and therapeutic approaches based on those tests. We also need to harmonize the regulatory approaches toward the transmission, storage and processing of large amounts of complex data, and to harmonize the guidelines and regulations for the conduct of genetic research across international borders.

For all involved — health care professional, regulator, reimburser — education regarding the understanding and interpretation of broad pharmacogenetic information needs to be expanded and intensified.

The integration of information technology, data management, and analytics to enable better understanding of the volumes of information being generated needs to be funded by either public or private sectors or, more likely, partnerships of both.

Finally, although there are clearly "believers", data need to be generated to either confirm the beliefs and convince the non-believers, or perhaps put the thing to rest.

Can anyone make money developing technology and approaches to personalized medicine? At least some venture capitalists think so. Several billion dollars have been pumped into the field in this decade. The investment flow peaked in 2001 when more than $800 million flowed into personalized medicine. Pharmaceutical companies are increasing their investments in the area — some more cautiously than others.

There are some new products and therapeutic products being delivered to the marketplace, and the co-development of accompanying diagnostics is expanding. That, in turn, is leading to an increase in laboratory services. In contrast, it is likely there will be more value-based and bundled pricing of products in the future. I also believe that evidence-based medicine will ultimately diminish pharmaceutical sales forces in the future.

Where might this lead? The economic model we have all been operating with for decades in this industry will likely change. Reimbursement models will change and pre-emptive (preventative) care will be covered because personalized medicine will have demonstrated cost savings and a better standard of care.

The blockbuster model of drug development will have changed to something more like a portfolio of products, each individually addressing segments of defined markets. Profiling of both new chemical entities and the targeted populations will be better, clearer and more extensive.

If all of that is true, the economic motivation will then be present to induce full pharmaceutical company participation.

So far, so good. But, while personalized medicine offers significant hope and opportunity, these things need to be in place to provide true benefit: funding (both investment and reimbursement), altered regulatory approaches, significantly enhanced IT capacity, industry, regulatory, legislative, professional commitment, and data. So the jury is still out; we ultimately need data.

15.2.2 Critical Path Initiative

In 2004, the FDA issued a report entitled, "Innovation — Stagnation". This report highlighted the FDA's concern about the lack of progress in drug development, what they perceived to be an innovation gap. They illustrated the report with the same type of information we presented in the first session in this course. Two years later, they published the Critical Path Opportunity List. That list specifically identified several target areas that, if effectively addressed, would result in improvements in the drug development process.

The list included biomarker development, streamlining clinical trials, bioinformatics and manufacturing. They also looked to combat emerging infections and bioterrorism and to develop therapies for children and adolescents.

The central thesis of the Critical Path Initiative was that society's commitment to investment in R&D for the purpose of improving the drug development process has been lacking. It cited a huge investment on the part of private and public sectors in basic research and in funding drug development, but a minor investment in development tools that might assist all areas of drug development.

In the FDA's words, the Critical Path Initiative is "a serious attempt to bring attention and focus to the need for targeted scientific efforts to modernize the techniques and methods used to evaluate the safety, efficacy and quality of medical products as they move from product selection and design to mass manufacture".

In a fairly simple-minded view, there are relatively few choices in terms of increasing the pipeline flow:

- ↑↑ Investment in R&D
- ↑↑ Candidates into the clinic
- ↓↓ Failure rate

One of these would be huge; two or three would result in massive change. One caution, though, is that simply pouring more money into R&D alone is not the answer; it must be effective money.

One thought (See Figure 15-1) is to go all the way back to discovery. Improve discovery . . . improve the delivery of drugs. If you can widen the funnel (theoretically at least) without changing other rates, you can increase the overall asset pool. However, we come back to our old friends — the resource constraints: manpower, budget, etc.

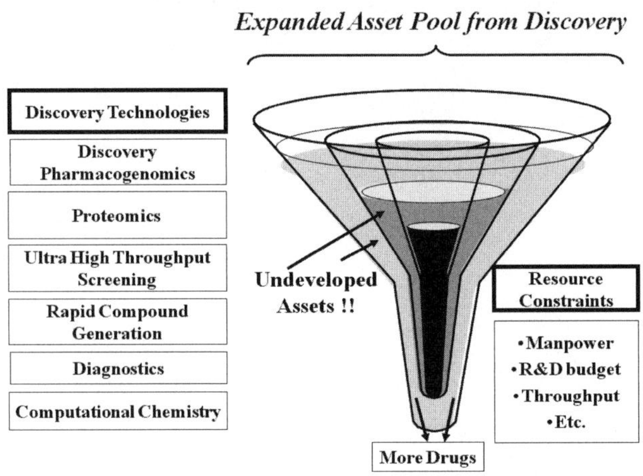

Figure 15-1. Improving Discovery Yields More Drugs

The FDA came up with some potential solutions to the Critical Path problem. First, they suggest utilizing new scientific knowledge to improve the medical product development process. Next, they suggest developing robust applied research programs into critical path scientific areas, hoping that will lead to new generalizable knowledge. They also want to strengthen academic bases for critical path disciplines and to intensify the FDA's involvement in critical path research and standards development. The FDA has defined the critical path as beginning in the prototype design (for devices) and mid-discovery (for drugs) stages and continuing through product approval.

Along the horizontal axis of the critical path (Figure 15-2), they define three additional dimensions — what the FDA calls 1) assessment of safety (looking at predictive markers of toxicity; how can one predict whether a potential product will be harmful?); 2) proof of efficacy (looking at predictors of benefit; how does one determine if a potential product will have medical benefit?); and 3) industrialization (looking at high quality manufacturability; how can one manufacture a product at commercial scale with consistently high quality?).

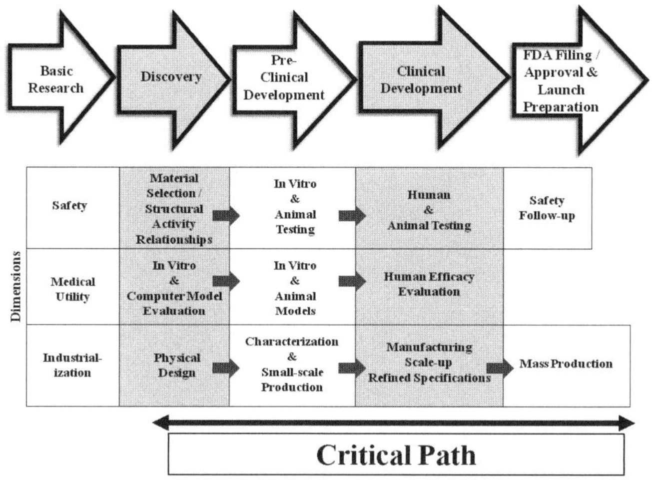

Figure 15-2. Critical Path: Multiple Dimensions

We mentioned a little while ago that the FDA's initial report was issued in 2004 and the opportunities list was issued in 2006. Other Critical Path-type initiatives have also been put in place since that time.

The Critical Path Institute (www.-path.org) was founded in 2005 in Arizona seeded by money primarily from the industry. It has, to date, functioned primarily as a think tank — holding conferences and discussion groups. Relatively few publications have emanated from those sessions. The Department of Health and Human Services has also put in place a personalized health care initiative, and the NIH has developed its Roadmap initiative.

The EU, not to be outdone, has created the Innovative Medicines Initiative (IMI) and, in a joint venture with industry, has funded it to the tune of two billion Euros (yes, with a "B"). (http://www.imi.europa.eu/index_en.html)

So, how might this play out?

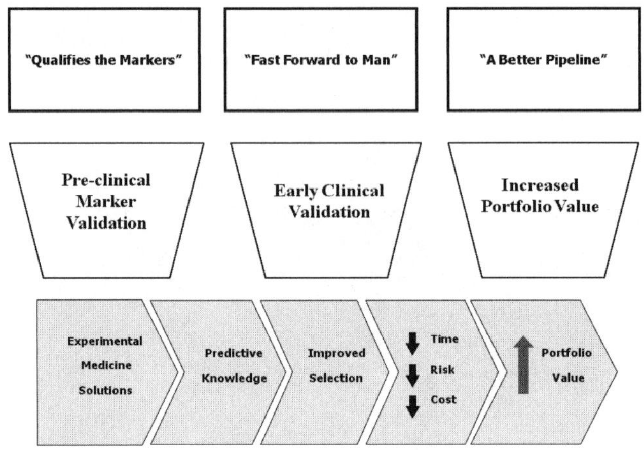

Figure 15-3. Improved Productivity in Drug Discovery

Figure 15-3 illustrates some potential opportunities and outcomes. If pre-clinical markers can be validated and can serve predictive functions for both safety and efficacy, then there are opportunities for moving quickly into man and biomarkers may serve as early go-no go signposts. If that is in place, there are chances that pipelines will be composed of higher-value opportunities, since they would have — once they got to that stage — a higher likelihood of success.

Ultimately, a program would require less time, have less risk at the points where expenditure of money is highest, and ultimately cost less. The sum total would be an increase in portfolio value.

As a corollary to the comments above, it is important — focusing on the clinical trials section of development — that the trial process, from preparation through execution, be streamlined. Options other than "mega trials" must be available and be useful in discriminating between successful and unsuccessful products.

If all of that were to come to pass, a new development paradigm begins to emerge. What I call "characterized" populations of patients (known to have certain characteristics or genetic makeup) would be engaged to study specific questions. This would enable smaller and more focused clinical studies that could be accomplished for less money in a shorter time frame. Ultimately, the process would lead to faster approvals.

I think the next few years will be exciting times to be in clinical research. However, the final outcomes aren't quite clear.

Over the next few years, the FDA plans to issue guidances such as end of Phase 2a structures, multiple endpoint studies, enrichment designs, non-inferiority designs and adaptive designs. And that leads us to the next topic: Adaptive Clinical Trials.

15.2.3 Adaptive Clinical Trials

Adaptive clinical trials are emerging as potential solutions to the issues we have been discussing — lengthening time lines, increasing costs and lack of predictive reliability in the drug development process.

It is a topic that seems to have arisen because of a series of fortuitous circumstances. Numerous groups — drug companies, regulators, academic investigators and statisticians — have become interested in the concept and potential implementation of these strategies. This has coincided with a re-birth of interest in Bayesian statistical approaches.

What are **adaptive clinical trials**? An adaptive trial design is one "that allows modifications to some aspects of the trial after its initiation without undermining the validity and integrity of the trial". (Chang, 2005)

Some potential benefits of adaptive trial designs include shorter and less expensive studies, more benefit / less harm to subjects, higher enrollment in terms of lower screen failure rates (ultimately fewer patients would enter the trials), and potentially enhanced compliance by physicians and patients. In addition, such designs offer the possibility of delivering safer and more targeted drugs to the market with faster regulatory approvals.

While being attractive in terms of their benefits, adaptive designs are operationally challenging.

To be effectively executed, the trials require electronic data capture with rapid review and an interactive voice response system to implement real-time randomization changes. Supplies must be deliverable in a just-in-time process that includes accurate labeling, packaging and distribution. The latter process only works with electronic inventory control (think Wal-Mart and Dell).

Table 15-5. Adaptive Clinical Trials: Required Ancillary Technologies

- Electronic Data Capture (EDC)
- Interactive Voice Response System (IVRS) or Interactive Web Response System (IWRS)
- Just-in-time clinical supply
    - Labeling
    - Packaging
    - Distribution
- Electronic inventory control
- Real-time site performance assessment systems
- Super organization and excellent execution

In addition real-time site performance assessment systems — tabulating screening, enrollment, visits, data processing, etc. must be in place. Finally, execution of these studies in an effective manner requires absolutely superb organization and flawless execution. They are not for the faint-hearted. However, if done well, they can be truly amazing.

Unfortunately, to provide a full appreciation of the details and execution of adaptive designs would require a long and detailed explanation. Given the constraints of the book, we won't be able to go into detail. However, as a very superficial overview, the designs enable the determination of patient outcomes as the data become available and, as the data are analyzed essentially in real time, modifications can be made during the conduct of the trial.

Some of the modifications that might be made include: re-estimation of sample sizes, early stopping due to efficacy or futility based on small numbers of patients, randomization that is adapted to the real-time data analyses, and discontinuation of treatment groups with inferior results and a re-randomization to other groups.

In February 2010, the FDA published its first draft guidance for Adaptive Design Clinical Trials. Stay tuned.

15.2.4 Bioinformatics

15.2.4.1 Modeling and Simulation

There are several opportunities in the bioinformatics area. The first involves modeling and simulation. The idea is to develop qualitatively and quantitatively predictive models of diseases that are able to incorporate the natural history of the disease, whatever biomarkers are known, and the range of effects of various interventions, as well as basic pharmacokinetic and pharmacodynamic information.

These models then enable "what-if" scenarios, looking at the results of various interventions including new drugs or therapeutic approaches. Certain of these models, particularly asthma and diabetes, appear to be reasonably robust.

15.2.4.2 Data Standards for Integration

Data standards, as anyone involved in data integration will affirm, are absolutely crucial to efficient data integration and data analysis. So many attempts to integrate databases over the past decade have required "one-off" solutions — a newly-created solution for each new integration.

The challenges here are almost mind-boggling; they stretch from earliest discovery through to post-market and involve safety and efficacy. Some (of the many) examples include:

Pre-Market

- Electronic submission
- Tracking processes
- Review processes

Post-Market
- Electronic adverse event reporting
- Database management

Quality
- Manufacturing regulation
- Inspection tracking

Scientific computing / computational science
- Academic laboratories
- Commercial laboratories
- Computational scientists

Serious issues remain in this area because of the requirements for data storage and computational power. Nevertheless, solutions to these issues will pay massive dividends.

15.2.4.3 Electronic Data Capture

Electronic data capture represents the acquisition of data at the site of generation and direct deposition into a database without further manual transcription. EDC systems are designed primarily for the collection of clinical data in an electronic format.

In order to be effective, these systems require a straightforward, intuitive graphical user interface for data entry. Behind the data entry format must be a validation system that permits real-time data validation (including data values and data types). On top of the system must be useful tools that enable data analysis and data reporting.

Such systems have been trumpeted for years as being the salvation for clinical trial data entry enabling the final elimination of paper records. To quote the old philosopher, "We ain't there yet!" Systems available today, however, are so much better than older systems and are very close to enable seamless timely data entry, analysis and reporting.

HOWEVER, the humans still cause problems. The biggest problems are at clinical trial sites. Physicians, study coordinators and other site personnel have generally been resistant to using these systems. Even today, this continues to be an issue. Nevertheless, I think this, again, is an area of great opportunity.

15.2.4.4 Electronic Health Records (EHR)

Electronic Health Records (EHR) is a topic about which I am quite passionate, but we don't have the space for all my ranting and raving. A survey published in 2008 reported that 4% of physicians had a fully functional electronic records system. (DesRoches, 2008) The same group surveyed US hospitals and in 2009 reported that only 1.5% had a comprehensive electronic records system. (Jha, et al. 2009)

I believe EHR will significantly reduce medical treatment errors and improve the standard of health care. While being cognizant of the potential security and privacy issues, I believe that portability of health information in a very mobile population would be quite helpful. Although older physicians — those of my generation — might dispute this statement, I believe EHR would actually improve physician efficiency and, likely, effectiveness.

I firmly believe there are profound opportunities in the area of clinical trial conduct and product development that remain untapped until widespread EHR are available.

Figures 15-4 and 15-5 illustrate, at the most superficial level, a system currently in place in the Harvard Partners medical practice. They have attempted to integrate the full medical record system including graphic information (such as EKGs, chest x-rays, scans, etc.), laboratory data, and history and physical examination data information into something they call the Clinical Research Chart.

The Clinical Research Chart (Figure 15-5) enables an integrated analysis of data regarding an individual patient at the current time as well as longitudinally through the patient's lifetime. Furthermore, the data can be vertically integrated, enabling analyses at a population level that might facilitate delivery of care within their own practice setting and, further, can be integrated with additional information to enhance the drug discovery and drug development processes.

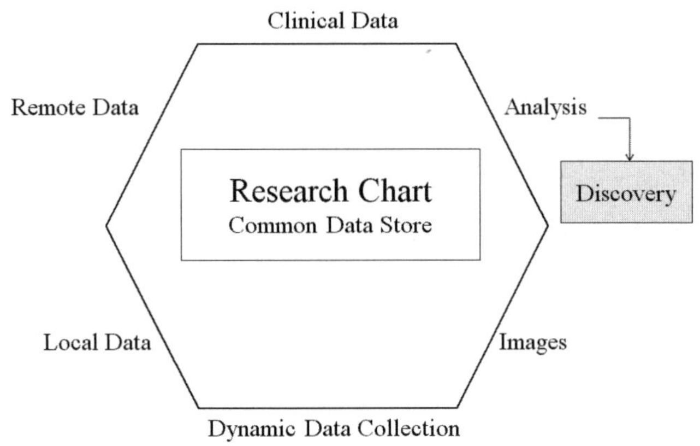

Figure 15-4. Harvard Partners: Clinical Data Integration

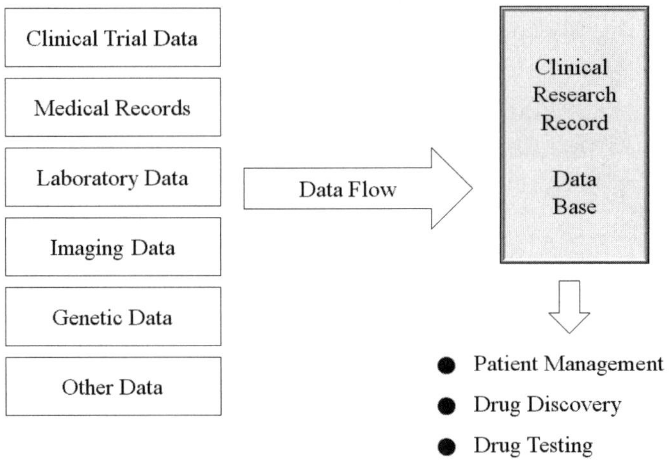

Figure 15-5. Harvard Partners: Clinical Research Chart

While this setup is real and in effect today, it represents a single practice — albeit large. There is a long way to go.

15.3 Focusing on the Future

Four topics don't do justice to the myriad of activities within the drug development arena that will translate into future changes. Additional topics of significant importance that we just don't have time in the book to touch upon include: further development of quantitative disease models, co-development of drug-diagnostic products, nanotechnology, indication-specific development — different paradigms for different diseases (such as pain, cancer, rheumatoid arthritis, etc.), and continued advances in the area of information technology, particularly in the areas of biomarkers and genetics.

References

DesRoches CM, Campbell EG, Rao SR, et al. Electronic Health Records in Ambulatory Care — A National Survey of Physicians. New Engl J Med (2008) 359:50-60.

Jha AK, DesRoches CM, Campbell EG, et al. Use of Electronic Health Records in U.S. Hospitals. New Engl J Med. (2009) 360:1628-38.

Summary and Conclusion

It has been my privilege to share my thoughts with you about many of the important issues in clinical development. Remembering that what we do in this industry impacts the lives of millions of people around the world, the responsibility for thoughtful, considered, ethical, efficient, and useful health care product development has never been weightier. It is an important business. It is all about our ultimate customer — the patient. So, do it well.

Comments and suggestions are welcome.

Barney King
Macnas Consulting International, Inc.
www.theclinicaltrialexpert.com

About the Author

Dr. King is President of Macnas Consulting International, Inc. Dr. King has nearly 25 years of experience managing business and development operations in the pharmaceutical, medical device, biotechnology and CRO (Contract Research Organization) industries.

Macnas specializes in strategic product development planning and execution for pharmaceutical, biotechnology and medical device companies and in due diligence consulting for venture capital, investment, and pharmaceutical firms. Dr. King's breadth and depth of experience are particularly valuable to smaller companies facing significant strategic product development decisions.

In addition to multiple projects on behalf of Macnas with a broad range of companies and venture capital firms addressing portfolio assessments, development programs, and strategic partnering, Dr. King's operational assignments at Macnas on behalf of various clients have included serving as President and CEO, Head of Development, Vice President, Clinical Affairs, Vice President Regulatory Affairs, and Medical Monitor.

Dr. King's previous industry positions have included: Senior Vice President, Development at Kemia, a biotechnology company; President and CEO of Predict, Inc., a bioinformatics company; Vice President and General Manager for San Diego Operations for PAREXEL, a Contract Research Organization, and Head of PAREXEL's Emerging Growth Opportunities Group; Executive Vice President for Biological Sciences and Medical Affairs at Trega Biosciences, a combinatorial chemistry company; President of Segenix, a spinout of Advanced Tissue Sciences;

and Vice President, Research and Development, Advanced Tissue Science, a biotechnology company. Dr. King also served as Vice President, Worldwide Medical and Regulatory Affairs for ConvaTec, a BristolMyers Squibb subsidiary, and Director of Clinical Investigation at SmithKline.

Dr. King graduated *cum laude* from the University of Notre Dame, majoring in English and Preprofessional Studies, and received his MD from the Ohio State University College of Medicine. Following completion of residency training in Medicine and Pathology at Ohio State and his Cardiology fellowship training at the Mt. Sinai Medical Center in New York, he was an Assistant Professor in Medicine (Cardiology) and Physiology at New York Medical College. He is board certified in Cardiology and Internal Medicine. He received his MBA from the Wharton School of the University of Pennsylvania.

Index

D

E

M

N

O

T